The
structure
of arithmetic

The Appleton-Century Mathematics Series

Raymond W. Brink and John M. H. Olmsted, Editors

HOWARD E. CAMPBELL

University of Idaho

The

structure

of arithmetic

APPLETON-CENTURY-CROFTS
EDUCATIONAL DIVISION
New York **MEREDITH CORPORATION**

Dedicated to the memory of

Professor Cyrus Colton MacDuffee

PREFACE

This text is designed especially for prospective and practicing elementary school teachers. In general, the material follows the recommendations of the Panel on Teacher Training of the Committee on the Undergraduate Program in Mathematics (CUPM) of the Mathematical Association of America. It is for a course in Mathematics and not for a course in arithmetic or teaching methods. It is assumed that the reader can perform the operations of elementary arithmetic and computation, but no other mathematical knowledge is assumed. In particular, no high school or college Mathematics courses are prerequisite.

It is not intended that the material presented be identical with what the elementary school teacher will teach to her (or his) own pupils. It is, however, chosen with the aim of giving the teacher a thorough understanding of the material she will teach, and also some understanding of parts of the mathematics that her pupils will study in higher grades. In order to do a really effective job of teaching mathematics with confidence, one should know more about the subject than one teaches, and should also know something about the material of later courses.

It is not suggested that the techniques of presentation in this text should be the same as those used in grade school. The most effective ways of teaching adults are not necessarily the most effective ways of teaching children.

The actual development of the structure of the number system, which begins in Chapter 3, is more mathematical than most other texts designed for elementary teachers. The book is also less nebulous, less computational, and more concise than most of the others. We use the definition, theorem, proof format with detailed explanations and examples. Most instructors will not want to cover all of the proofs in a course. However, essentially all of the proofs are included for completeness so that they are readily available to students.

At first glance, some readers may feel that the present development is too difficult for the students for whom it is intended. However, the text in its evolving form has been used since the summer of 1965 in all sections of the required Mathematics sequence in the elementary eductaion major at the University of Idaho with a high degree of success. Included in these sections were both prospective and practicing elementary teachers, most of whom had little mathematical training and many of whom came to the course with considerable fear or lack of interest in Mathematics. The fears were almost all eliminated in the first part of the course and many of the students commented toward its end that for the first time they really understood and enjoyed Mathematics.

All of the instructors who have taught the course have been surprised at the amount and depth of the mathematics that they were able to teach using preliminary versions of this text. The theoretical development as presented has made the material easier rather than more difficult to understand.

The emphasis is on the understanding of ideas and not on the ability to manipulate. In particular, a student should be able to state precisely any of the definitions or theorems (not necessarily using the same words as the text, but necessarily saying the same thing). More than that, he should understand thoroughly what the words are actually saying. A student should also be able to prove things whose proofs are not of a long or involved nature.

When reading a mathematics book a student should have pencil and paper at hand. He should read very carefully because every word, symbol, and punctuation mark is important. At any difficult spot one should spend a reasonable amount of time trying to understand the troublesome thing and then proceed even if it is not understood. This same technique should be continued until several such points are not understood or the assignment has been completed. One should then start again and reread the material. Usually it will become clear at a second or third reading, and in the meantime the student continues to make progress on the material. The reason this technique is so valuable is that later material nearly always helps to clarify earlier parts.

The help of Donna P. Cinkosky, Gail A. Jervik, Melita E. Vest, and Sandra L. Hazen in typing various versions of the text and of Karen J. Davidson and Bonnie B. Anderson in typing the final manuscript is gratefully acknowledged. Thanks are also extended to Paul F. Dierker, Erol Barbut, Judith S. Gates, and the students in Math 15 and 16 at the University of Idaho for their helpful ideas and discussions. In addition I would like to express my appreciation to Karen J. Van Houten for checking the answers to the exercises and to Susan D. Bohlander for helping with the proofs. Finally, I thank Professor Raymond W. Brink and the other editors of Appleton-Century-Crofts for all their help in the preparation of this book.

Moscow, Idaho   H.E.C.

CONTENTS

The
structure
of arithmetic

chapter 1

Preliminaries

1.1 Introduction

Mathematics consists essentially of *undefined terms*, *definitions*, *axioms*, *deductive reasoning*, and *theorems*. In order to have an understanding of any kind of Mathematics, one should understand these notions.

Mathematical definitions are made for convenience, and, of course, a mathematician is at liberty to define a term in any way he chooses, with due regard to accepted mathematical usage. Naturally, it is important that a definition be stated carefully and precisely. It should be realized, however, that one cannot continually define words in terms of simpler words forever; there must be some starting point at which the terms used are actually *undefined terms*. One is not completely in the dark about the meaning of undefined terms, however, because one gets some knowledge of them from their use in defining the other terms.

If one looks up a word in a dictionary, he will, of course, find that a word is defined in terms of other words. If he then looks up each of these words and continues to look up each word that is used in their definitions, he will eventually come back to terms that he had already looked up. Hence, these words were essentially used to define themselves. We stress again that since words cannot continually be defined in terms of simpler words, there must be some undefined words as starting points.

Our formal mathematical development will not begin until Chapter 3, where it will be pointed out that the undefined terms that we will start with are the terms "set" and "element." In our formal development, definitions will be numbered, and labeled as definitions. Before Chapter 3 our treatment will be informal and will attempt to give the reader some familiarity with general mathematical terminology and logical reasoning.

A **theorem** is a proposition that one proves from other propositions using **deductive reasoning**. (We shall discuss deductive reasoning in Chapter 2.) These other propositions may themselves be theorems which in turn were proved from still other propositions. As in the case of definitions, it is clear that propositions cannot always be obtained from simpler propositions; there must be some starting point. There must be some propositions which one assumes, propositions that one does not prove from others. These basic propositions are called **axioms**. The axioms are the basic assumptions of the particular kind of mathematics involved. Using the undefined terms and the terms that one has defined by means of them, one proceeds from the axioms to obtain more propositions called theorems.

A good example of the use of axioms and theorems is encountered in ordinary high school plane geometry where one is exposed to axioms such as "through two distinct points, there is one and only one line," and theorems such as "the diagonals of a parallelogram bisect each other."

1.2   The mathematical use of certain words

In order to understand the precise meaning of spoken or written mathematics, and also in order to write or speak precisely in mathematics, one should understand the mathematical meaning of a number of words and combinations of words. Four of these are listed below with descriptions of their meanings.

1. There exists: there is at least one.
2. For every: for each
3. Unique: exactly one; no more and no less than one.
4. Some: at least one.

By way of illustration, consider the line segment AB (Figure 1.2.1). We shall use the notation AB to mean the entire line segment including the end points A and B. In this example, let the length of AB be 2 inches, and let C be the midpoint of AB. We will analyze some statements, which are either true or false, about this segment.

Figure 1.2.1

Statement 1. For every point P on segment AB, there exists a unique point P' (which depends on P) on AB whose distance from P is 1 inch.

This statement says that no matter what point P is chosen on AB, there is exactly one (no more and no less than one) other point P' also on AB whose distance from P is 1 inch. The statement is false, however, because if P is taken at C, there will be two points (and hence not a unique point) on AB whose distance from P is 1 inch, namely, points A and B.

Let us consider some additional statements.

Statement 2. For every point P different from C on AB, there exists a unique point P' (which depends on P) on AB whose distance from P is 1 inch.

This statement is true because as long as P is not taken at C there will be exactly one point P' (no more and no less) which is 1 inch from P.

Statement 3. For every point P on segment AB, there exists a point P' on AB whose distance from P is 1 inch.

Statement 3 is true because it does not require P' to be unique.

Statement 4. For every point P on AC, there exists a point P' on AB which is to the right of P.

This is true; in fact, for every point P on AC there exists an infinite number of points on AB which are to the right of P.

Statement 5. For every point P on AC, there exists a point P' (which depends on P) on AC which is to the right of P.

This is false since there is one exception. No such P' exists if P is the point C on AC. We can change Statement 5 to make a different statement which is true, however, if we start the sentence with " For every point P to the left of C on AC ...."

Statement 6. There exist some points on AB which are not to the right of A.

This statement says that there is at least one point on AB which is not to the right of A. The statement is true since A is such a point.

Statement 7. Some points on AB which are to the right of C are to the right of A.

Statement 7 is also true, because all of the points to the right of C on AB are to the right of A, and hence at least one of them is.

Exercises

Determine which of the following statements are true and which are false. Give reasons for your answers.

1. For every living person P, there exists a unique person whom P will marry;
2. For every unmarried living person P, there exists a unique person whom P will marry;

3. For every unmarried living person *P*, there exists a person whom *P* will marry;

4. For every person *P* who will marry at some future time, there exists a unique person whom *P* will marry;

5. For every person *P* who will marry at some future time, there exists a person whom *P* will marry;

6. There exists a unique number *x* such that $2x$ is larger than 4;

7. There exists a number *x* such that $x^2 = 4$ (x^2 means *x* times *x*);

8. There exists a unique positive number *x* such that $x^2 = 4$;

9. There exists a unique number *x* such that $xy = yx = y$ for every number *y* (the symbol xy means *x* times *y*);

10. Some numbers which are larger than 8 are larger than 5;

11. Some points on a line segment *AB* are equidistant from *A* and *B*;

12. Some numbers are between 5 and 7;

13. For every number *n*, there exists a number, *n'*, larger than *n*;

1.3   The mathematical use of "and," "or," and "not"

We will now consider the use of three other words in mathematics, namely, "and," "or," and "not." Their use is in forming statements from other statements. The kind of statements we shall be considering are those that are either true or false (but not both true and false).

The use of "and"

A new statement can be obtained by putting the connective "and" between two statements. The only time a statement so constructed is true is when both of its parts are true.

Example 1. $3 > 5$ *and* there exists a number $x \neq 3$ ($>$ means "is greater than").

Example 2. Mars is a planet *and* there are no people 20 feet tall.

The statement in the first example is false, because one of its parts ($3 > 5$) is false, while the statement in the second example is true because both of its parts are true.

The use of "or"

A new statement can be obtained by putting the connective "or" between two statements. In mathematics, the "or" is used differently from the way it is usually used in ordinary non-mathematical conversation such as "Bill will go *or* George will go," where the interpretation is that either Bill will go or George will go but not both. If the "or" used in the given statement were the mathematical "or," the interpretation would be: Either Bill will go or George will go, or they will both go. The mathematical "or" is the "inclusive" rather than the "exclusive or." It means one or the other or both. The only time a statement obtained by putting "or" between two given statements is false is when both of the given statements are false.

Example 3. There exists a number $x \neq 3$ *or* $7 > 4$. This statement is true.

Example 4. $3 > 5$ *or* there exists a number $x \neq 3$. This one is also true.

Example 5. Mars is not a planet *or* there are people 20 feet tall. This statement is false because both of its parts are false.

The use of "not"

If S is a statement, then the statement "not S" can be obtained from it. The statement "not S" is called the **negation** of the statement S. If one takes a statement and puts "not" in front of it, the result is not always a nice-sounding statement or even a grammatically correct one. For example, putting "not" in front of the statement "Mars is a planet" does not produce a nice-sounding statement or a grammatically correct one, so one would write instead "Mars is not a planet." If a given statement S is true, then the statement "not S" is false. However, if the given statement S is false, then the statement "not S" is true.

There may be more than one way of expressing the negation of a statement. For example, consider "All cows are black." Here the negation could be stated as "Not all cows are black" or as "Some cows are not black" or as "There exists at least one cow that is not black." Notice that although the statement "Some cows are brown" contradicts the given statement, it is not the negation of the given statement. Also, the statement "All cows are not black" is not the negation, because it means "There are no black cows." Notice that the statement "All cows are black" is false and that the statement "There are no black cows" is also false, so that one of them is not the negation of the other.

1.4   Truth tables and negations

If A is a statement and B is a statement, there are four cases for the truth or falsehood of their combination, namely:

        Case 1. They are both true.

        Case 2. They are both false.

        Case 3. A is true while B is false.

        Case 4. A is false while B is true.

For each of these possibilities, let us consider the "truth value" (that is, the truth or falsehood) of the statement *A and B*. We obtain:

Case 1. True.
Case 2. False.
Case 3. False.
Case 4. False.

Using T for true and F for false, we can conveniently summarise this information as follows:

A	B	A and B
T	T	T
F	F	F
T	F	F
F	T	F

This **truth table** shows how the "truth value" of a statement *A and B* depends on the truth values of the statements from which it is constructed. Similarly, we can get the truth table that shows how the truth value of the statement *A or B* depends on the truth values of statements from which it is built. We obtain:

A	B	A or B
T	T	T
F	F	F
T	F	T
F	T	T

We can, in fact, conveniently indicate in a single table how the truth value of each of the above statements depends on the truth values of its parts, as follows:

A	B	A and B	A or B
T	T	T	T
F	F	F	F
T	F	F	T
F	T	F	T

For the sake of convenience, we will often use *S'* to denote the negation of any statement *S*. With this notation, we have the following truth table for the relative truth values of *S* and *S'*.

S	S'
T	F
F	T

Using the same notation (A' for the negation of statement A, and B' for the negation of statement B) we obtain the following table, which shows how the truth values of various statements depend on the truth values of statement A and statement B.

A	B	A and B	A or B	A'	B'	A' and B'	A' or B'	A' and B
T	T	T	T	F	F	F	F	F
F	F	F	F	T	T	T	T	F
T	F	F	T	F	T	F	T	F
F	T	F	T	T	F	F	T	T

Notice that the table shows that the negation of the statement A *and* B is A' *or* B' because it shows that for all of the four possible combinations of the truth values of statement A and statement B, the truth values of A' *or* B' are opposite those of A *and* B. The table also shows that A' *and* B' is not the negation of A *and* B, because they have the same truth values for two of the four cases. Similarly we see from the table that the negation of A *or* B is A' *and* B'.

In a sense, the connective "and" and the connective "or" are opposites of each other, because we replace one by the other when we obtain the negation of a statement of the form A *and* B or a statement of the form A *or* B. (Of course when doing this we also replace A by its negation and B by its negation.)

Example 1. The negation of

"No cows are asleep and all people are tired."

is obtained by replacing the "and" by "or" and by changing each of the statements "no cows are asleep" and "all people are tired" to its negation. We obtain

"Some cows are asleep or some people are not tired."

Example 2. The negation of

"Some cats are not hot or some goats eat."

is

"All cats are hot and no goats eat."

Exercises

1. How does the above table show that neither A' *and* B' nor A' *and* B is the negation of A *and* B?

2. Make additional columns for A *and* B', A' *or* B, and A *or* B' in the above table, and state three other things that the resulting table shows.

Write the statements that are the negations of the following statements, and decide, when you can, whether each statement and each negation is true or false.

3. Mars is a planet and all cows are black;

4. Mars is a planet or all cows are black;

5. Not all cows are black and Mars is a planet;

6. Bill is wearing a red tie and George is wearing a tie;

7. Bill is not wearing a tie or Jim is tired;

8. Some people are hot or all cows are black;

9. No cows are white and some people are simple;

10. Not all cows are tall and no people are wide;

11. Some cows are not simple and not all people are problems;

12. Some corn is not sweet or some cows are not problems.

chapter 2

Reasoning and implication

2.1 Validity of reasoning

We shall be concerned with the task of proceeding by logical reasoning from a set of statements called the **hypothesis** to a resulting statement called the **conclusion**. We shall often need to answer the question: If the hypothesis is assumed to be true, does its truth *force* the conclusion to be true? If it does, the reasoning is described as **valid**; otherwise, it is described as **invalid**. In deciding whether the reasoning is valid or invalid, we will not be concerned with the actual truth or falsehood of the hypothesis or the conclusion but only with whether the assumed truth of the hypothesis forces the truth of the conclusion, that is, whether the hypothesis makes the conclusion inescapable.

Actually, the hypothesis may be true or false and so may the conclusion. In fact, it is possible to have each of the following cases:

Case	Hypothesis	Conclusion	Reasoning
1	True	True	Valid
2	True	True	Invalid
3	True	False	Invalid
4	False	True	Valid
5	False	True	Invalid
6	False	False	Valid
7	False	False	Invalid

However, by the meaning of valid reasoning it is not possible to have the hypothesis true, the reasoning valid, and the conclusion false. That is, *if the hypothesis is true and the reasoning is valid, then the conclusion must be true.*

Example 1.

Hypothesis: All dogs are animals and all hounds are animals.

Conclusion: All hounds are dogs.

Here the hypothesis and conclusion are both true, but the reasoning is invalid! The fact that both dogs and hounds are animals does not, in itself, force all hounds to be dogs.

We could get a "picture" or diagram of the situation in the above example by representing animals, dogs, and hounds as the points inside closed curves. The hypothesis requires the dog region and the hound region to be inside the animal region but does not force the hound region to be inside the dog region. The diagram can be drawn as in Figure 2.1.1.

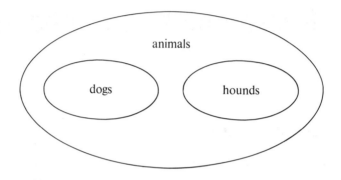

Figure 2.1.1

It is usually convenient to draw one of these **Euler diagrams** (named after the famous eighteenth century Swiss mathematician Leonard Euler) from the given hypothesis to help decide on the validity of the reasoning. One must be careful, though, because one is not often forced to draw the diagram in a unique way. It should be emphasized that the diagram is only an aid to deciding validity and is not necessarily equivalent in some sense to the hypothesis. For example, the hypothesis of Example 1 does not force the hound region and dog region not to overlap. The diagram could be drawn as in Figure 2.1.2.

If the conclusion had been "some dogs are hounds," the reasoning would also be invalid because the hypothesis does not force the diagram to be drawn in this second way.

Example 2.

Hypothesis: Some cows are brown.

Conclusions: (a) Some cows are not brown.

(b) Some brown things are cows.

Here a diagram could look like either Figure 2.1.3 or Figure 2.1.4.

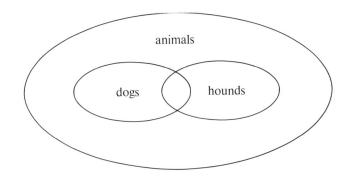

Figure 2.1.2

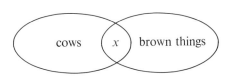 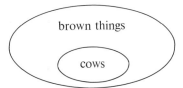

Figure 2.1.3   *Figure 2.1.4*

The x is to indicate that we are forced to have an overlapping of the cow region and the brown things region. For conclusion (a) the reasoning is invalid because the hypothesis itself does not prevent all cows from being brown. The reasoning for conclusion (b), however, is valid because the hypothesis forces the conclusion.

Notice that in this example the hypothesis is true and both conclusions are true. It is important to realize that truth is a property of statements and validity is a property of reasoning. Truth and validity do not have the same meaning.

Example 3.

Hypothesis: All weeds are plants.

Conclusion: Some plants are not weeds.

The reasoning is invalid because the hypothesis does not force the conclusion. It does not prevent all plants from being weeds. It could be that plants and weeds are the same thing. So in the diagram (Figure 2.1.5) the region for plants could be the same as the region for weeds. In this example, the hypothesis is true and the conclusion is true, but the reasoning is invalid.

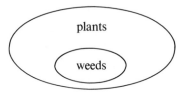

Figure 2.1.5

Example 4.
Hypothesis: All dogs are animals and all hounds are dogs.
Conclusion: All hounds are animals.
Here the reasoning is valid. The hypothesis forces the hound region inside the dog region and the dog region inside the animal region and, hence, forces the hound region to be inside the animal region (Figure 2.1.6).

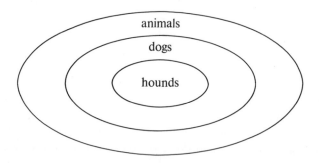

Figure 2.1.6

Example 5.
Hypothesis: Some blocks are rocks and all rocks are smooth things.
Conclusion: Some blocks are smooth things.
An Euler diagram for this hypothesis could appear as in Figure 2.1.7.
In this case the hypothesis is false, the conclusion is true, and the reasoning is valid. The blocks that are rocks are smooth things. The x in the diagram indicates that we are forced to have the block region overlap the rock region.

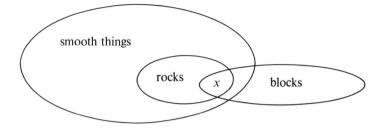

Figure 2.1.7

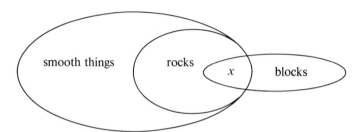

Figure 2.1.8

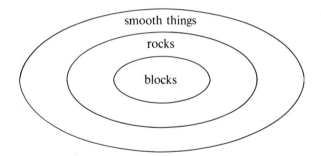

Figure 2.1.9

If the conclusion had been: "Some blocks that are not rocks are smooth things," then the reasoning would have been invalid, because the hypothesis does not force this conclusion. Figure 2.1.8. illustrates this. The hypothesis also does not force some blocks that are not rocks to be not smooth things either. Hence, if the conclusion were "Some blocks that are not rocks are not smooth things," the reasoning would still be invalid. A diagram representing the given hypothesis as being true but representing this conclusion as false is Figure 2.1.9.

Exercises

For each hypothesis draw a diagram and use it to help decide if the reasoning for each conclusion is valid or invalid. For each case where the reasoning is invalid, draw an additional diagram if your first one does not already show that it represents the hypothesis as being true but the conclusion false. For Hypotheses 5 and 6, give your opinion of the truth of the hypothesis and each conclusion.

1. Hypothesis: All Communists are integrationists. Peter Smith is an integrationist.
   Conclusion: Peter Smith is a Communist.

2. Hypothesis: All Communists are integrationists and Khrushchev is a Communist.
   Conclusion: Khrushchev is an integrationist.

3. Hypothesis: Some swizzles are not blobs and some nungs are swizzles;
   Conclusion: (a) Some nungs are not blobs.
   (b) Some nungs are blobs.
   (c) Some blobs are swizzles.
   (d) Some swizzles are nungs.
   (e) Some blobs are not swizzles.

4. Hypothesis: All nins are grips and all grips are toves.
   Conclusion: (a) All toves are nins.
   (b) All nins are toves.

5. Hypothesis: Some cows are animals and some animals have hair.
   Conclusion: (a) Some cows have hair.
   (b) Some animals are cows.
   (c) Some animals are not cows.

6. Hypothesis: All animals are remembered and some cows are animals.
   Conclusion: (a) Some cows are remembered.
   (b) Some cows are not remembered.
   (c) Some cows that are not animals are remembered.
   (d) Some cows that are not animals are not remembered.

2.2 Implication

We have been considering reasoning expressed in the form:

Hypothesis: A

Conclusion: B

where A and B are statements. For our purposes, when the reasoning is valid we will say that A **implies** B (often written as $A \Rightarrow B$) or say that the statement "A implies B" is true. When the reasoning is invalid, we will say that A does not imply B (often written as $A \nRightarrow B$) or say that the statement "A implies B"

is false. A statement of the form "*A* implies *B*" is called an **implication**. There are a number of other ways which are frequently used to express the implication "*A* implies *B*." They are:

(1) If *A*, then *B*.
(2) *B* is a necessary condition for *A*.
(3) *A* is a sufficient condition for *B*.
(4) *B* follows from *A*.
(5) *A* only if *B*.
(6) *B* if *A*.

Naturally these statements will have the same truth value as the statement "*A* implies *B*." Consider for example the following implication:

If George falls in the water and he is wearing only
a conventional swimsuit, then he will get wet.

This could equivalently be expressed in each of the following ways:

(1) (George falls in the water and he is wearing only a conventional swimsuit) implies (George will get wet).

(2) (George will get wet) is a necessary condition for (George falls in the water and he is wearing only a conventional swimsuit).

(3) (George falls in the water and he is wearing only a conventional swimsuit) is a sufficient condition for (George will get wet).

In order to illustrate implications further, let us consider the following situation. Suppose that there are two cork balls, one colored white and one colored green, in a box, along with one black steel ball. We will consider that the cork balls are light (in weight) and the steel ball is heavy. Assume that George reaches into the box and takes out a ball. Let us consider the following implications relative to this situation:

(1) If George gets a white ball, then he gets a light ball.
(2) If George gets a light ball, then he gets a white ball.
(3) If George gets a heavy ball, then he gets a black ball.
(4) If George gets a black ball, then he gets a heavy ball.

Notice that each of these implications except the second one is true, whereas the second one is false. It is extremely important to realise that an implication of the form $A \Rightarrow B$ may be true when the implication $B \Rightarrow A$ is false, or vice versa. The implication $B \Rightarrow A$ is called the **converse of** $A \Rightarrow B$. That is, the converse of an implication is formed by interchanging the hypothesis and the conclusion.

It may also be the case that the implications $A \Rightarrow B$ and $B \Rightarrow A$ are both true (or both false). As illustrations, let us consider the four implications above. Implication (2) is the converse of implication (1). In this case, the implication is true while its converse is false. Also, implication (1) is the converse of implication (2), and this is a case where the implication is false while its converse is true. Similarly, implications (3) and (4) are converses of each other, and both of them are true.

There is a great tendency among people to confuse an implication and its converse. Suppose we consider, as another example, the following implication:

> If Jim stole the cookies, then he will not be able to
> keep from blushing when his mother enters the room.

The converse is:

> If Jim is not able to keep from blushing when his
> mother enters the room, then Jim stole the cookies.

If we accept the first of these statements as being true, we should be careful not to consider the second statement to be necessarily true. We should be careful not to punish Jim without getting further information: perhaps Jim is blushing because he just kissed Suzie Cromwell and not because he stole the cookies. The distinction between an implication and its converse cannot be overemphasized, both in mathematics and in everyday occurrences.

Another implication that people often confuse with $A \Rightarrow B$ is $A' \Rightarrow B'$ which is called the **inverse** of $A \Rightarrow B$. As an example, consider:

> If Jim goes to the store, he will buy some apples.

The inverse is:

> If Jim does not go to the store, he will not buy some apples.

If the first of these implications is true, then the second one may or may not be true. Jim may not buy apples if he does not go to the store, but on the other hand he may not go to the store and still may buy some apples from a door-to-door peddler. In general, an implication and its inverse may or may not have the same truth value.

We have seen that for an implication $A \Rightarrow B$ the converse $B \Rightarrow A$ and the inverse $A' \Rightarrow B'$ may or may not have the same truth value as $A \Rightarrow B$. There is still a third implication related to $A \Rightarrow B$, namely, $B' \Rightarrow A'$, which is called the **contrapositive** of $A \Rightarrow B$. We will consider the contrapositive in detail in Section 2.3 and see that an implication and its contrapositive always have the same truth value.

Another important type of statement that one meets is the **double implication**:

> *A* if and only if *B*

which means

> (*A* if *B*) and (*A* only if *B*)

or

$$B \Rightarrow A \text{ and } A \Rightarrow B,$$

and it is often symbolized by

$$A \Leftrightarrow B.$$

As we have seen previously $A \Rightarrow B$ can be expressed as *A is a sufficient condition for B*, while $B \Rightarrow A$ can be expressed as *A is a necessary condition for B*. Hence,

$$A \text{ if and only if } B$$

can be expressed as:

A is a necessary and sufficient condition for B.

When A is a necessary and sufficient condition for B, each of the statements A and B implies the other and they are called **equivalent** statements. It follows from the definition of valid reasoning that equivalent statements have the same truth value.

As an illustration, notice that in the situation above involving George and the balls, the statements "George gets a black ball" and "George gets a heavy ball" are equivalent because each implies the other. However, the statements "George gets a green ball" and "George gets a light ball" are not equivalent, because the first of these implies the second but the second does not imply the first.

Exercises

Using George and the box of balls, decide the truth value of each of the following implications:

1. If George does not get a white ball, then he does not get a light ball.
2. If George does not get a light ball, then he does not get a white ball.
3. If George does not get a heavy ball, then he does not get a black ball.
4. If George does not get a black ball, then he does not get a heavy ball.
5. George gets a black ball, if and only if he gets a heavy ball.
6. George gets a heavy ball, if and only if he gets a black ball.
7. George gets a light ball, if and only if he gets a green ball.
8. George gets a white ball, if and only if he gets a light ball.

If A and B are statements and the implication $A \Rightarrow B$ is true, decide whether each of the implications 9–11 (a) must be true, (b) must be false, or (c) may be either true or false.

9. $B \Rightarrow A$.   10. $A' \Rightarrow B'$,   11. $B' \Rightarrow A'$.

State (a) the converse, (b) the inverse, and (c) the contrapositive of each of the following implications. In Exercises 18–20 also determine the truth values of the given implication and each of its three related implications.

12. If Bill goes, then Jim will go.
13. If Jim does not go, then Bill will not go.
14. If Jim goes, then Bill will go.
15. If $x = 2$, then $y = 7$.
16. If $x \neq 2$, then $y \neq 7$.
17. If $x = 3$, then $y \neq 5$.
18. If $2x = 4$, then $x = 2$.
19. If $x \neq 2$, then $2x \neq 4$.
20. If $x + y = 8$, then $x = 2$ and $y = 6$.

2.3   The contrapositive

In this section we will show that an implication $A \Rightarrow B$ and its contrapositive $B' \Rightarrow A'$ are equivalent and hence always have the same truth value. That is, we will show that $A \Rightarrow B$ if and only if $B' \Rightarrow A'$. Before we do this, however, recall that the meaning of valid reasoning requires that if the reasoning is valid it cannot be the case that the hypothesis A is true and the conclusion B is false. In other words, if $A \Rightarrow B$ is true, then the statement "A is true and B is false" must be false. Of course, this requires that the negation of the latter statement be true. That is, it requires that the statement "A is false or B is true" be true.

What we have just proved is that:

(1)   *If $A \Rightarrow B$, then A is false or B is true.*

But to say A is false is the same as saying A' is true, so

(2)   *If $A \Rightarrow B$, then A' is true or B is true.*

In order to show that

$$A \Rightarrow B \text{ if and only if } B' \Rightarrow A',$$

we will need to prove both of the following:
   (a) If $A \Rightarrow B$, then $B' \Rightarrow A'$,
   (b) If $B' \Rightarrow A'$, then $A \Rightarrow B$.

Proof of (a) If $A \Rightarrow B$, then $B' \Rightarrow A'$.

We will show that if $A \Rightarrow B$, then the truth of B' forces the truth of A'. By (2), if $A \Rightarrow B$, the A' is true or B is true. Hence (when $A \Rightarrow B$), if B' is true then A' must be true. In other words, if $A \Rightarrow B$, then $B' \Rightarrow A'$.

Proof of (b)   If $B' \Rightarrow A'$, then $A \Rightarrow B$.

We have just shown in (a) that when a first statement implies a second statement, then the negation of the second statement implies the negation of the first statement. Applying this to $B' \Rightarrow A'$, we get $A \Rightarrow B$; so if $B' \Rightarrow A'$, then $A \Rightarrow B$.

The major use of the fact that $A \Rightarrow B$ is equivalent to $B' \Rightarrow A'$ is that if we wish to prove that a first statement implies a second, we may do so by proving instead the contrapositive, namely, that the negation of the second statement implies the negation of the first statement. It often happens that the contrapositive is easier to deal with than the implication itself.

Consider the following example. Suppose that a company with 900 employees, 500 being men over 25 years of age, and 200 having no dependents, wants to determine whether all of their male workers over the age of 25 have dependents. That is, they want to check the truth, for their situation, of the implication:

> If a man is over the age of 25, then he has dependents.

To do this, they will have to check the information for 500 men. However, it is easier to check the contrapositive, namely:

> If an employee has no dependents, then he is not a man over the age of 25,

because they will then have to check information for only the 200 people who do not have dependents, to see if none of them is a man over the age of 25.

Consider another example. Suppose that George, Jim, and Bill are sitting blindfolded and necktieless and that Mary clips a bow tie on each of them, saying that each of the ties is either red or green but at least one is red. They then remove their blindfolds so that each can see the color of the other two ties but not the color of his own. Each of the fellows is supposed to deduce his own tie color and tell the other people in the room what it is. George sees a red tie on Bill and a green tie on Jim.

A few moments pass. Relative to this situation, let us prove the following statement:

> If Bill does not say that he has a red tie, then George has a red tie.

It is not difficult to prove this statement by proving its contrapositive, namely,

> If George does not have a red tie, then Bill will
> say that he has a red tie.

This is true because if George does not have a red tie, Bill will see that neither George nor Jim has a red tie, and since Bill knows that at least one of the ties is red, Bill will say that he has a red tie.

Notice how much easier it is in this case to work with the contrapositive statement. It might be instructive to try to prove the statement directly without using the contrapositive. You will find that you are naturally forced into the contrapositive.

Exercise

1. Suppose that the preceding game is played as before except that only Bill and George remove their blindfolds, while Jim keeps his on. A number of minutes pass but no one speaks.

If you are Jim, how can you use deductive reasoning to deduce that you have a red tie? (We will assume in this exercise that both George and Bill are rather intelligent people and that Mary is honest.)

(*Hint*: Prove the contrapositive of the implication: if neither Bill nor George claims to have a red tie, then Jim has a red tie, That is, prove the following: If Jim does not have a red tie, then either Bill or George will claim to have a red tie. To prove this, assume, that Jim's tie is not red and consider the following two possible cases:

(1) Bill's tie is not red.

(2) Bill's tie is red.)

chapter 3

Sets

3.1 Sets, elements, subsets, and set equality

Sets are of fundamental importance in all of mathematics. We will use them as a foundation for the number system.

If one tries to define the term **set,** he comes up with such things as " a set is a collection of objects," or " a set is a bunch of elements," or something similar. If one then tries to define " collection " or " bunch," he gets a similar sort of statement using some other synonym, and so he cannot really define every one of these words by using other simpler words. We accept *set* as an **undefined term**. It is a specific illustration of our previous discussion about the necessity of undefined terms.

We will refer later to some of the following examples of sets:

(1) The set S_1 of points on a line segment AB.
(2) The set S_2 of people in the United States.
(3) The set S_3 of blue-eyed cats owned by people named Egbert.
(4) The set S_4 of whole numbers from 2 to 16 inclusive.
(5) The set S_5 of elephants who manufacture tire pumps.
(6) The set S_6 of people in the State of Idaho.
(7) The set S_7 of blue-eyed cats.
(8) The set S_8 of whole numbers from 11 to 19 inclusive.
(9) The set S_9 of all cats each of which has 17 tails.
(10) The set S_{10} that consists of the numbers 2,3,4.
(11) The set S_{11} of whole numbers larger than 12.
(12) The set S_{12} of whole numbers from 2 to 19 inclusive.
(13) The set S_{13} of whole numbers from 11 to 16 inclusive.

If we try to define the term **element** of a set, we get into the same sort of difficulty as when trying to define the term *set*. We might have tried to say the elements of a set are the " objects of the set." Here again, the words used help us understand the meaning of the term but still leave us with the word " object " to define. We will be content to leave the term **element** of a set undefined.

We will thus begin our formal development with two undefined terms, namely, **set** and **element** of a set. It is not appropriate in this text to give the formal axioms of set theory. It is appropriate however, to mention that, as far as our study of sets is concerned, the formal axioms essentially assume that sets exist and that the ways we will use to get new sets from given sets, such as union and intersection and the method of Section 3.5, actually give us things that are sets.

In the foregoing examples, the elements of the set S_1 are the points on a segment AB, the elements of S_2 are the persons in the United States, the elements of S_4 are the whole numbers from 2 to 16, etc. Note that presumably sets S_5 and S_9 have no elements.

We will write $a \in S$ to indicate that "a is an element of set S" and some-times say, in this case, that "a belongs to S" or "a is in S." For example, $5 \in S_4$ and $11 \in S_8$. The statement "a is not an element of S" will be written as $a \notin S$. For example, $22 \notin S_8$.

For convenience, we will frequently use curly brackets to indicate the set whose elements are written inside them. For example, $\{1, 2, 3\}$ will denote the set whose elements are 1, 2, and 3, while $S_{13} = \{11, 12, 13, 14, 15, 16\}$. Similarly, $\{1, 2, 3, \ldots, n\}$ means the set whose elements are the whole numbers from 1 to n. The symbol "$\ldots$" is used in mathematical notation to mean " and so on until we get to " or, if there is no number following the three dots, they mean " and so on without stopping." Of course, this symbol should be used only when there is no doubt about what it represents.

We will now make some formal definitions.

Definition 3.1.1 (*Definition of subset*) *If every element of a set S is an element of a set T, S is said to be a **subset** of T.*

This definition can be stated more concisely with symbols as: A set S is a *subset* of a set T if $(x \in S) \Rightarrow (x \in T)$.

We will use the notation $S \subset T$ to denote that S is a subset of T. $S \subset T$ is often read "S is contained in T." $S \not\subset T$ means that S is not a subset of T. Clearly every set is a subset of itself, since $(x \in S) \Rightarrow (x \in S)$.

Definition 3.1.2. (*Definition of equality of sets*) *Two sets S and T are called* **equal***, and we write S = T, if and only if every element of each set is an element of the other.*

Note that if $S = T$, then $S \subset T$ and $T \subset S$. Also note that if $S \subset T$ and $T \subset S$, then $S = T$. We will write $S \neq T$ to indicate that sets S and T are not equal.

Examples of equal sets:

(1) $S_8 = \{11, 12, 13, 14, 15, 16, 17, 18, 19\}$.
(2) Every set is equal to itself.
(3) S_{11} = the set of whole numbers greater than or equal to 13.
(4) $\{a, b, c, d\} = \{c, a, d, b\}$.

Definition 3.1.3. (*Definition of proper subset*) *If $S \subset T$ and $S \neq T$, then S is called a **proper subset** of T.*

Examples:

(1) S_6 is a subset of S_2, that is, $S_6 \subset S_2$. S_6 is a proper subset of S_2.
(2) S_{13} is a subset of S_4, that is, $S_{13} \subset S_4$. S_{13} is a proper subset of S_4.

(3) $S_3 \subset S_7$. If a person knows of a living blue-eyed cat not owned by a person named Egbert, he can state that S_3 is a proper subset of S_7.

(4) Every set is a subset of itself, but no set is a proper subset of itself.

We will now state and prove our first theorem.

Theorem 3.1.1. *Set equality has the following three properties (where R, S, and T are sets):*

(a) $S = S$ (*that is, every set is equal to itself*). (*This is called the **reflexive** property of set equality.*)

(b) *If $S = T$, then $T = S$.* (*This is called the **symmetric** property of set equality.*)

(c) *If $S = T$ and $T = R$, then $S = R$ (that is, sets equal to the same set are equal to each other).* (*This is called the **transitive** property of set equality.*)

Proof.

(a) S has the same elements that S does, so, by definition of set equality, $S = S$.

(b) If S and T have the same elements, then T and S have the same elements; so, by definition of set equality, if $S = T$, then $T = S$.

(c) If S and T have the same elements and T and R have the same elements then S and R have the same elements, and so if $S = T$ and $T = R$, then $S = R$.

One should be careful to distinguish between the phrases "is a subset of" and "is an element of." In particular, one should distinguish between a set and its elements. This is especially the case for a set with a single element. The set containing the element 1 is not the same thing as 1, that is, $\{1\}$ is not the same as 1 because $\{1\}$ is a set (whose only element is 1) but 1 is a number and not a set. Note that 1 is an element of $\{1\}$ and $\{1\}$ is a subset of $\{1\}$, but 1 is not a subset of $\{1\}$, and $\{1\}$ is not an element of $\{1\}$; that is, $1 \in \{1\}$ and $\{1\} \subset \{1\}$ but $1 \not\subset \{1\}$ and $\{1\} \notin \{1\}$.

It is possible, however, for a set to be an element of another set. For example, $\{a, b\} \in \{\{a, b\}, c\}$ but $\{a, b\} \not\subseteq \{\{a, b\}, c\}$. However, it is true that $\{\{a, b\}\} \subseteq \{\{a, b\}, c\}$.

How does $S = \{a, b\}$ differ from $T = \{\{a, b\}\}$? They certainly do not have the same elements. The set S described here contains two elements, namely, a and b, whereas T contains only one element, namely, the set $\{a, b\}$. Note also that $\{a\} \subseteq \{a, b, c\}$ but $\{a\} \notin \{a, b, c\}$, while $a \in \{a, b, c\}$ but $a \not\subseteq \{a, b, c\}$.

3.2   The null set

Note that both S_5 and S_9 contain no elements. They are equal by the theorem we will now prove.

Theorem 3.2.1. *If S is a set with no elements and T is a set with no elements, then $S = T$; that is, there is a unique set with no elements.*

Proof. If S contains no elements and T contains no elements, then the sets S and T have the same elements, and so by Definition 3.1.2, $S = T$.

Definitions 3.2.1. (*Definition of empty set or null set*) *The unique set containing no elements is called the* **empty set** *or the* **null set***.*

We will use the usual symbol $\varnothing$ to denote the empty set. Note that $\{\varnothing\} \neq \varnothing$ because $\varnothing$ contains no elements while $\{\varnothing\}$ contains an element, namely the set $\varnothing$.

Theorem 3.2.2. *The null set $\varnothing$ is a subset of every set T.*

Proof. To prove this, we will prove the contrapositive of: If $a \in \varnothing$, then $a \in T$; that is, we will prove that if $a \notin T$, then $a \notin \varnothing$. This latter statement is true since $\varnothing$ contains no elements.

3.3   Union, intersection, and disjoint sets

There are several important ways in which one can get a set from given sets.

Definition 3.3.1. *The* **union** *of sets S and T is the set consisting of the elements that are in either S or in T (or in both).*

We use the notation $S \cup T$ for the union of sets S and T.

Examples of set union:

(1) $S_4 \cup S_8$ is the set of whole numbers from 2 to 19; that is, $S_4 \cup S_8 = S_{12} = \{2, 3, 4, \ldots, 19\}$.

(2) $S_7 \cup S_2$ is the set whose elements are all blue-eyed cats and the people in the United States.

(3) $S_4 \cup S_{13} = S_4$.

(4) $S_{12} \cup S_5 = S_{12}$.

(5) $S_{11} \cup \emptyset = S_{11}$.

Something to be careful of is illustrated as follows:

Let $S = \{a, b, c,\}$ and $T = \{1, 2\}$. Then $S \cup T = \{a, b, c, 1, 2\}$. But $\{S, T\} = \{\{a, b, c\}, \{1, 2\}\} \neq S \cup T$.

One can see that $S \cup T$ contains five elements while $\{S, T\}$ contains exactly two elements, namely, the set S and the set T.

Definition 3.3.2. *The **intersection** of sets S and T is the set of elements that are in both S and T.*

We use the notation $S \cap T$ for the intersection of sets S and T.

To help remember Definition 3.3.2, recall that two distinct nonparallel lines intersect in the point that belongs to both lines.

If we let the set of points on one of the lines in Figure 3.3.1 be S and the set of points on the other line be T, then $S \cap T = \{P\}$.

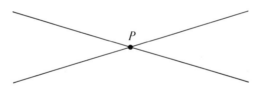

Figure 3.3.1

Examples of set intersection:

(1) $S_4 \cap S_8$ is the set of numbers from 11 to 16; that is,
$$S_4 \cap S_8 = S_{13} = \{11, 12, 13, 14, 15, 16\}.$$

(2) $S_{10} \cap S_4 = S_{10}$.

(3) $S_{10} \cap S_{11} = \emptyset$.

(4) $S_{11} \cap S_{13} = \{13, 14, 15, 16\}$.

(5) $S_7 \cap S_2 = \emptyset$.

Definition 3.3.3. *The sets S and T are called **disjoint** if and only if $S \cap T = \emptyset$; that is, if and only if they have no elements in common. Two or more sets are called disjoint if no two of them have an element in common.*

Examples of disjoint sets:

(1) S_{10} and S_{11} are disjoint, since $S_{10} \cap S_{11} = \emptyset$.

(2) S_7 and S_8 are disjoint.

Exercises

1. Find all the subsets of each of the following sets, being careful to use the correct notation for each of them. (*Hint*: The number of subsets you need to find in each case is (a) 1, (b) 2, (c) 4, (d) 8, (e) 4.)
   (a) $\emptyset$;   (b) $\{7\}$;   (c) $\{1, 2\}$;   (d) $\{a, b, c\}$;   (e) $\{a, \{b, c\}\}$.

2. Why does $\{1, 2, 3, 4, 5\} = \{4, 1, 5, 3, 2\}$?

3. Referring to the sets S_1 to S_{12} of Section 3.1, find
   (a) $S_{11} \cup S_{12}$;   (b) $S_{11} \cap S_{12}$;   (c) $S_3 \cup S_{10}$;   (d) $S_3 \cap S_{10}$;
   (e) $S_{11} \cup \emptyset$;   (f) $S_{11} \cap \emptyset$.

4. If $S = \{a, b, c\}$, $T = \{b, c, d, e\}$, and $R = \{a, b, d, g\}$, find
   (a) $S \cap T$;   (b) $S \cup T$;   (c) $T \cap R$;   (d) $(S \cup T) \cap R$;
   (e) $T \cup \emptyset$;   (f) $(S \cap T) \cap R$;

5. Is $\{0\} = \emptyset$? Why?

6. Is $\emptyset \subset \{0\}$? Why?

7. Why is not $\{a\} \in \{a, b, c\}$?

8. Why is $\{a\} \in \{\{a\}, b, c\}$?

9. If S and T are sets, why is $S \cap T = T \cap S$ and why is $S \cup T = T \cup S$?

10. Find (a) $(S_4 \cap S_{13}) \cup S_8$ and (b) $(S_4 \cup S_8) \cap (S_{13} \cup S_8)$, referring to the sets S_1 to S_{12} of Section 3.1.

3.4   Euler diagrams, differences, and complements

Diagrams like those used in Chapter 2 to help decide the validity of certain reasoning are also useful in getting "pictures" of various properties and relations of sets. As we have mentioned previously, these are called Euler diagrams. Another name for them is Venn diagrams. If we represent the elements of a set as the points inside or on a closed curve, then various set relations can be pictured as in Figure 3.4.1.

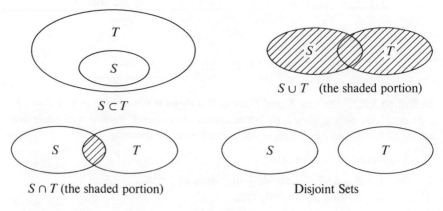

Figure 3.4.1

Definition 3.4.1. *If S and T are sets, the **difference** S − T is the set of elements of S that are not elements of T.*

A diagram for this is Figure 3.4.2, where the shaded portion is $S - T$.
For example if $S = \{a, b, c, d\}$ and $T = \{b, d, g, h, p\}$ then $S - T = \{a, c\}$ and $T - S = \{g, h, p\}$.

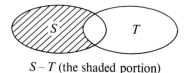

$S - T$ (the shaded portion)

Figure 3.4.2

Definition 3.4.2. *If a set A is a subset of a set S, then the **complement** of A in S is the set S − A, that is, the set of elements of S that are not elements of A.*

For example, the complement of $\{a, b, c\}$ in $\{a, b, c, d, h\}$ is $\{d, h\}$.
When there is no doubt about the set S to which we are referring, we will write A' in place of $S - A$ for the complement of A in S.
A diagram of a complement is Figure 3.4.3, where the shaded portion is A'.

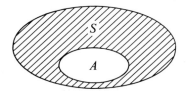

Complement of A in S (the shaded portion)

Figure 3.4.3

3.5   More on set notation and set construction

Another notation that is often used to specify a set is illustrated by:

$$\{x \,|\, x \text{ is alive and } x \text{ has 6 legs}\}.$$

This is read as "The set of all x such that x is alive and x has 6 legs." Of course, this is the set of all living things with 6 legs. An additional example is:

$$S \cup T = \{x \,|\, x \in S \text{ or } x \in T\},$$

which is read as "The set of all x such that x is an element of S or x is an element of T."

Another way of constructing sets from other sets is described as follows: Let S be a set and obtain a new set from it by saying that the new set is the set of all elements of S that have a certain property. Of course, a set obtained in this way is a subset of the given set S.

The set of all elements of a set S that have a certain property can be expressed as:

$$\{x \in S \mid x \text{ has that certain property}\}.$$

Because this is a way of constructing sets, this notation is often called **set builder notation.**

For example, if we let S be the set of people in the United States,

$$\{x \in S \mid x \text{ owns an airplane}\}$$

stands for the set of people in the United States who own an airplane. Other examples of this method of set specification are:

(1) $S - T = \{x \in S \mid x \notin T\}.$
(2) $S \cap T = \{x \in S \mid x \in T\} = \{x \mid x \in S \text{ and } x \in T\}.$
(3) $\emptyset = \{x \in S \mid x \notin S\}.$

One reads this notation just as it is written except that " { " is read " the set of all" and the vertical line " $\mid$ " is read "such that." For example, $\{x \in S \mid x \notin T\}$ is read as "The set of all x in S such that x is not in T."

Exercises

1. Draw a diagram similar to Figure 3.5.1 and shade it to represent each of the following sets:

(a) $S \cap T$;
(b) $R \cup (S \cap T)$;
(c) $(R \cup S) \cap (R \cup T)$;
(d) $R \cap (S \cup T)$;
(e) $(R \cap S) \cup (R \cap T)$;
(f) $R \cap (S \cap T)$;
(g) $(R \cap S) \cap T$;
(h) $(R \cap S) - T$;
(i) $(R \cup T) - S$;
(j)   the complement of $(S \cap T)$ in R.

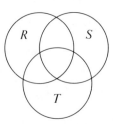

Figure 3.5.1

2. Let $S = \{1, 2, 3, \ldots\}$, the set of all positive whole numbers, and let T be the set of all people in the State of Idaho, and let

$A = \{x \in S \mid 2 < x < 7\}, \qquad B = \{x \in S \mid 4 < x < 12\}. \qquad$ Find:

(a) $A - B$;   (b) $B - A$;
(c) $S - A$;   (d) $A - S$;
(e) $T - S$;   (f) $S - T$;
(g) The complement of A in B;   (h) The complement of A in S.

3. Express the following sets in the correct notation (take S and T to be the sets of Exercise 2);

(a) The set of elements in S and in T but not in $S \cap T$;
(b) The set of all people in Idaho who are less than 21 years of age;
(c) The set of all people in Idaho who are tall and wear glasses;
(d) The set of all whole numbers from 17 to 92 inclusive;
(e) The set of positive even numbers;
(f) The set of positive whole numbers that are multiples of 5;
(g) The set of positive whole numbers greater than 18.

3.6   Additional properties of union and intersection of sets

It has already been pointed out in Exercise 9 of Section 3.3 that if S and T are sets, then $S \cup T = T \cup S$ and $S \cap T = T \cap S$. One can prove these set equalities directly from the definitions of union, intersection, and equality of sets: the elements that are in either S or T are the same as those that are in either T or S, and the elements that are in both S and T are those that are in both T and S; that is, $S \cup T = T \cup S$ and $S \cap T = T \cap S$.

We have proved the first and fifth parts of the following theorem.

Theorem 3.6.1. *If R, S, and T are sets, then:*

(1) $S \cup T = T \cup S$ *(This is called the **commutative** property of set union);*
(2) $(R \cup S) \cup T = R \cup (S \cup T)$ *(This is called the **associative** property of set union);*
(3) $R \cup (S \cap T) = (R \cup S) \cap (R \cup T)$⎫   *(Set union is distributive on both*
(4) $(S \cap T) \cup R = (S \cup R) \cap (T \cup R)$⎭   *sides over set intersection);*
(5) $S \cap T = T \cap S$ *(Set intersection is commutative);*
(6) $(R \cap S) \cap T = R \cap (S \cap T)$ *(Set intersection is associative);*
(7) $R \cap (S \cup T) = (R \cap S) \cup (R \cap T)$⎫   *(Set intersection is distributive*
(8) $(S \cup T) \cap R = (S \cap R) \cup (T \cap R)$⎭   *on both sides over set union).*

In order to use Euler diagrams to help understand the steps in the proofs of parts (2), (3,) (6), and (7) of this theorem, we note that an Euler diagram showing all possible ways in which sets R, S, and T could overlap would be as Figure 3.6.1.

We shall leave the proofs of parts (2), (7), and (8) of Theorem 3.6.1 as exercises.

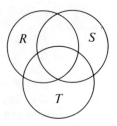

Figure 3.6.1

Proof of (6). $(R \cap S) \cap T = R \cap (S \cap T)$.

By the definitions of intersection and equality of sets,

$(R \cap S) \cap T = \{x \mid x \in R \cap S \text{ and } x \in T\} = \{x \mid x \in R \text{ and } x \in S \text{ and } x \in T\}$.

Similarly,

$R \cap (S \cap T) = \{x \mid x \in R \text{ and } x \in S \cap T\} = \{x \mid x \in R \text{ and } x \in S \text{ and } x \in T\}$.

Therefore $(R \cap S) \cap T = R \cap (S \cap T)$.

The significant Euler diagrams for the steps of the proof are shown in Figure 3.6.2.

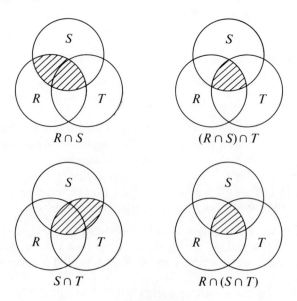

Figure 3.6.2

Proof of (3). $R \cup (S \cap T) = (R \cup S) \cap (R \cup T)$.

$R \cup (S \cap T) = \{x \mid x \in R \text{ or } x \in S \cap T\} = \{x \mid x \in R \text{ or } (x \in S \text{ and } x \in T)\}$
$(R \cup S) \cap (R \cup T) = \{x \mid (x \in R \text{ or } x \in S) \text{ and } (x \in R \text{ or } x \in T)\}$
$\qquad\qquad\qquad\qquad = \{x \mid x \in R \text{ or } (x \in S \text{ and } x \in T)\}.$

Therefore $R \cup (S \cap T) = (R \cup S) \cap (R \cup T)$.

The significant Euler Diagrams for the proof of (3) are shown in Figure 3.6.3.

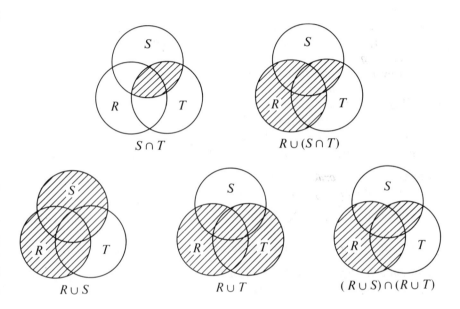

Figure 3.6.3

Proof of (4). $(S \cap T) \cup R = (S \cup R) \cap (T \cup R)$.
By (1), $(S \cap T) \cup R = R \cup (S \cap T)$ and so by (3) and (1),
$(S \cap T) \cup R = R \cup (S \cap T) = (R \cup S) \cap (R \cup T) = (S \cup R) \cap (T \cup R)$.
Therefore, $(S \cap T) \cup R = (S \cup R) \cap (T \cup R)$.

Exercises

Use techniques similar to the above to prove parts (2), (7), and (8) of Theorem 3.6.1. For (2) and (7), draw the significant Euler diagrams for the steps of your proofs.

3.7   The Cartesian product

The above Theorem 3.6.1 will be of use in showing that addition and multiplication of numbers have certain properties. Another concept which will help us in defining multiplication of numbers is the **Cartesian product** of two sets. Before we see what a Cartesian product is however, we must understand what is meant by an **ordered pair**.

According to the definition of equality of sets, $\{a, b\} = \{b, a\}$ because these sets have the same elements. Under what circumstances would $\{r, s\} = \{g, h\}$? This equality would hold if and only if $r = g$ and $s = h$ or $r = h$ and $s = g$.

However, if we are concerned not only about what the elements of a set are but also about the order in which the elements are taken, the set will be an **ordered set**. We shall be concerned here with ordered sets of two elements and will call them **ordered pairs.** Parentheses will be used to denote **ordered pairs.** If $a \neq b$, there are two different ordered pairs whose elements are a and b, namely, (a, b) and (b, a). The formal definition we will use is as follows.

Definition 3.7.1. *An **ordered pair** (s, t) is a set whose elements are s and t, where $(s, t) = (g, h)$ if and only if $s = g$ and $t = h$.*

For example, $(s, t) = (3, 4)$ if and only if $s = 3$ and $t = 4$. Also, $(1, 2) \neq (2, 1)$.

We are now able to define the **Cartesian product** $S \times T$ of sets S and T.

Definition 3.7.2. *The **Cartesian product** $S \times T$ of sets S and T is the set of all ordered pairs (s, t) where $s \in S$ and $t \in T$.*

Another way of expressing this definition is as follows:

$$S \times T = \{(s, t) \,|\, s \in S \text{ and } t \in T\}.$$

Example 1. Let $S = \{1, 2\}$ and $T = \{2, 3, 4\}$. Then the elements of the set $S \times T$ are $(1, 2)$, $(1, 3)$, $(1, 4)$, $(2, 2)$, $(2, 3)$, $(2, 4)$, hence in this case $S \times T = \{(1, 2), (1, 3), (1, 4), (2, 2), (2, 3), (2, 4)\}$.

Example 2. Let $S = \{4, 5\}$ and $T = S$. Then

$$S \times T = S \times S = \{(4, 4), (4, 5), (5, 5), (5, 4)\}.$$

Note that $(4, 5) \neq (5, 4)$ and that we have not required the elements of $S \times T$ to be expressed in any particular order, so for example we could also say $S \times T = S \times S = \{(5, 4), (4, 4), (5, 5), (4, 5)\}$.

Theorem 3.7.1. *If S is any set, then* $S \times \varnothing = \varnothing$ *and* $\varnothing \times S = \varnothing$.

Proof. Here we get $\varnothing$ for the Cartesian product, because there are no elements of the form (s, t) where $s \in S$ and $t \in \varnothing$ since there are no elements t such that $t \in \varnothing$. Similarly, $\varnothing \times S = \varnothing$.

We shall use Theorem 3.7.1 to prove that zero times a whole number is zero, in Chapter 4. The following theorem about Cartesian products will be of use in proving that multiplication of whole numbers is distributive over addition.

Theorem 3.7.2. *If S, G, and H are sets, then* $S \times (G \cup H) = (S \times G) \cup (S \times H)$.

Proof. By the definition of Cartesian product,

$$S \times (G \cup H) = \{(s, t) \mid s \in S \text{ and } t \in G \cup H\}.$$

But, by the definition of union, $t \in G \cup H$ can be expressed as $t \in G$ or $t \in H$, so that

$$S \times (G \cup H) = \{(s, t) \mid s \in S \text{ and } (t \in G \text{ or } t \in H)\}.$$

Also, $S \times G = \{(s, t) \mid s \in S \text{ and } t \in G\}$; $S \times H = \{(s, t) \mid s \in S \text{ and } t \in H\}$. Hence $(S \times G) \cup (S \times H) = \{(s, t) \mid s \in S \text{ and } (t \in G \text{ or } t \in H)\}$, which is the same thing we arrived at for $S \times (G \cup H)$. Therefore, $S \times (G \cup H) = (S \times G) \cup (S \times H)$.

Exercises

Let $S = \{1, 2, 3\}$, $T = \{a, b, c, d\}$, $R = \{1, 2, a, c\}$, $G = \{g, p\}$, $H = \{h\}$. Find:

1. $G \times S$;
2. $S \times G$;
3. $H \times G$;
4. $S \times T$;
5. $T \times S$;
6. $(S \times G) \times H$;
7. $S \times (G \times H)$;
8. $S \times S$;
9. $G \times G$.
10. (a) Find $R \cup T$ and then find $S \times (R \cup T)$.
    (b) Find $S \times R$ and $S \times T$ and then find $(S \times R) \cup (S \times T)$.
    (Note that you get the same set for (a) as for (b)—Theorem 3.7.2 guarantees this.)

3.8 One-to-one correspondence

Most of us have at least a vague idea of what a number is. One of our main purposes in this text is to clarify understanding of this fundamental concept at a suitable level of rigor. We still need one more important set concept in order to proceed, namely, the concept of **one-to-one correspondence.**

Consider the following situation. Suppose that a party, with both men and women, is held in a room containing a record player. Suppose further that every man is dancing with exactly one woman and that every woman is dancing with exactly one man. With the men and women "paired" in this way, we would know that the number of men is the same as the number of women in the room, that is, that the number of elements in the set of men in the room is equal to the number of elements in the set of women in the room. It would certainly not be necessary to count the people to know this.

Consider another room full of people and chairs, where each person is sitting on exactly one chair and each chair is being occupied by exactly one person. Here, similarly to the case above, with this kind of a "pairing" we would know that the number of people in the room is the same as the number of chairs in the room, that is, that the number of elements in the set of people in the room is equal to the number of elements in the set of chairs in the room.

These two examples illustrate the concept of **one-to-one correspondence** which we shall now define. We shall later use one-to-one-correspondences to assist in the definition of the number of elements in a set.

Definition 3.8.1. *If the elements of a set S can be paired with the elements of a set T in such a way that every element of S is paired with a unique element of T and every element of T is paired with a unique element of S, then the pairing is called a* **one-to-one correspondence** *between S and T.*

For convenience, the expression "one-to-one correspondence" is often written as "1–1 correspondence." According to Definition 3.8.1, the null set cannot be put in 1–1 correspondence with any set $S \neq \varnothing$. We will consider that it can be put in 1–1 correspondence with itself, however.

Example 1. Let $S = \{a, b, c\}$ and $T = \{p, q, r\}$. Then the pairing $a \leftrightarrow p$, $b \leftrightarrow q$, and $c \leftrightarrow r$ is a 1–1 correspondence between S and T. Of course, there are other 1–1 correspondences between these sets, for example, $a \leftrightarrow r$, $b \leftrightarrow q$, and $c \leftrightarrow p$ is another. Notice that $a \leftrightarrow p$, $b \leftrightarrow r$, $a \leftrightarrow r$, $c \leftrightarrow q$ is not a 1–1 correspondence between S and T, because neither a nor r is paired with a unique element.

Example 2. Let $S = \{a, b\}$ and $T = \{\text{boot, shoe, cat}\}$. If we pair a with one of the elements of T and then pair b with one of the elements of T, we will have an element of T left over. Hence these sets cannot be put in a 1–1 correspondence.

Example 3. Let $S = \{1, 2, 3, \ldots, n, \ldots\}$ and $T = \{2, 4, 6, \ldots, 2n, \ldots\}$. Then $1 \leftrightarrow 2$, $2 \leftrightarrow 4$, $3 \leftrightarrow 6$, $4 \leftrightarrow 8$, $\ldots$, $n \leftrightarrow 2n$, $\ldots$ is a 1–1 correspondence between S and T. Note that T is a proper subset of S, so that in this case a set is put in 1–1 correspondence with a proper subset of itself!

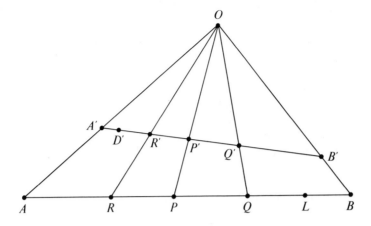

Figure 3.8.1

Example 4. The sets of points on two line segments of different (or the same) lengths can be put in 1–1 correspondence. In particular, in Figure 3.8.1 let AB and $A'B'$ be the segments and let O be the point of intersection of lines AA' and BB'. Let S be the set of points on segment AB, and let T be the set of points on segment $A'B'$.

Here the set of all lines through O which intersect segment AB gives us a pairing for a 1–1 correspondence between S and T; for example, $P \leftrightarrow P'$, $Q \leftrightarrow Q'$, $R \leftrightarrow R'$, $A \leftrightarrow A'$, $B \leftrightarrow B'$, etc.

Definition 3.8.2. *If there is a* 1–1 *correspondence between sets S and T, they are called* **equivalent sets**, *and we write* $S \sim T$.

For example, if $S = \{a, b, c, d\}$ and $T = \{1, 2, 3, 4\}$, then $S \sim T$. Also, as we saw in Example 3, the set of positive whole numbers is equivalent to the set of positive even numbers.

If S is any set, then $S \sim S$, because we can pair each element of S with itself to get a 1–1 correspondence. Also, if $S \sim T$, then $T \sim S$, because a 1–1 correspondence of sets S and T is a 1–1 correspondence of sets T and S. Thus equivalence of sets is reflexive and symmetric. Is it also transitive? Consider the following sets: $S = \{1, 2, 3\}$, $T = \{p, q, r\}$, $R = \{\text{cat, dog, boat}\}$. Then

$$
\begin{array}{ll}
1 \leftrightarrow p & p \leftrightarrow \text{cat} \\
2 \leftrightarrow q & q \leftrightarrow \text{dog} \\
3 \leftrightarrow r & r \leftrightarrow \text{boat}
\end{array}
$$

are 1–1 correspondences between S and T and between T and R. We can use these to get a 1–1 correspondence between S and R by using the elements of T as intermediaries as follows: $1 \leftrightarrow$ cat (the things paired with p), $2 \leftrightarrow$ dog (the things paired with q), and $3 \leftrightarrow$ boat (the things paired with r).

Similarly, if we have any sets S, T, and R such that $S \sim T$ and $T \sim R$, then there is a 1–1 correspondence between S and T, say $a \leftrightarrow a'$, $b \leftrightarrow b'$, $c \leftrightarrow c'$, ..., and one between T and R, say $a' \leftrightarrow a''$, $b' \leftrightarrow b''$, $c' \leftrightarrow c''$, ...

Again we can use the elements of T as intermediaries to get the 1–1 correspondence $a \leftrightarrow a''$, $b \leftrightarrow b''$, $c \leftrightarrow c''$, ..., between S and R. We summarize these results in the following theorem.

Theorem 3.8.1. *If R, S, and T are sets, then*

(1) $S \sim S$ (*Set equivalence is* **reflexive**);
(2) *If $S \sim T$, then $T \sim S$* (*Set equivalence is* **symmetric**);
(3) *If $S \sim T$ and $T \sim R$, then $S \sim R$* (*Set equivalence is* **transitive**);

Another result which will be of use to us in proving that multiplication of numbers is commutative and associative is the following theorem.

Theorem 3.8.2. *If R, S, and T are sets, then*

(1) $S \times T \sim T \times S$;
(2) $R \times (S \times T) \sim (R \times S) \times T$.

Proof. A 1–1 correspondence between $S \times T$ and $T \times S$ is the pairing that pairs every element (s, t), where $s \in S$ and $t \in T$, with the element (t, s) of $T \times S$. A pairing that is a 1–1 correspondence between $R \times (S \times T)$ and $(R \times S) \times T$ is the one that pairs every element $(r, (s, t))$ of $R \times (S \times T)$ with the element $((r, s), t)$ of $(R \times S) \times T$.

The next theorem will also be useful in Chapter 4.

Theorem 3.8.3. *If S and T are disjoint sets and S' and T' are also disjoint sets such that $S \sim S'$ and $T \sim T'$, then*

(1) $S \cup T \sim S' \cup T'$;
(2) $S \times T \sim S' \times T'$. (*This is true even if $S \cap T \neq \varnothing$.*)

Proof of (1). $S \cup T \sim S' \cup T'$.

Since $S \cap T = \varnothing$, every element of $S \cup T$ is either an element of S that is not in T or an element of T that is not in S. Similarly, every element of $S' \cup T'$ is either an element of S' that is not in T' or an element of T' that is not in S'. Because $S \sim S'$ and $T \sim T'$, there is a 1–1 correspondence between S and S' and a 1–1 correspondence between T and T'.

We obtain a 1–1 correspondence between $S \cup T$ and $S' \cup T'$ by pairing an element of $S \cup T$ that is in S with the same element of S' it is paired with in the first of the above-mentioned 1–1 correspondences, and by pairing an element of $S \cup T$ that is in T with the same element of T' it is paired with in the second of the above-mentioned 1–1 correspondences.

What we have done is just to combine a 1–1 correspondence of S and S' with a 1–1 correspondence of T and T' to get a 1–1 correspondence of $S \cup T$ and $S' \cup T'$.

Proof of (2). $S \times T \sim S' \times T'$. (Our proof holds even if $S \cap T \neq \varnothing$.)

Here again we can combine two 1–1 correspondences to get the one that we require. Each element (s, t) of $S \times T$ where $s \in S$ and $t \in T$ is simply paired with the element (s', t') of $S' \times T'$, where s' is the element of S' that is paired with s in the first of the 1–1 correspondences mentioned previously in the proof of (1), and where t' is the element of T' that is paired with t in the second of these 1–1 correspondences.

Exercises

1. Find all of the 1–1 correspondences between $S = \{a, b, c\}$ and $T = \{p, q, r\}$ (*Hint*: There are six.)

2. In Example 3, what element of S is paired with the element 234 of T and what element of T is paired with 196 of S?

3. In Example 4, how do you determine the point L' of $A'B'$ that is paired with the point L of AB, and how do you determine the point D of AB that is paired with the point D' of $A'B'$. Draw the diagram.

4. Find a 1–1 correspondence between the sets $S =$ the set of positive whole numbers, and $T =$ the set of whole numbers greater than 100. In your correspondence, what element of T is paired with an element n of S? What element of S is paired with an element h of T?

5. Let $R = \{1, 2\}$, $S = \{a, b, c\}$, and $T = \{x, y\}$, and write an explicit 1–1 correspondence between each of the following:
   (a) $R \times S$ and $S \times R$;
   (b) $(R \times S) \times T$ and $R \times (S \times T)$.

6. Let AB be a line segment and let C be a point between A and B. Show that the points on AB can be put in 1–1 correspondence with the points on AC. (*Hint*: Take a segment $A'B'$ above AB and set up, as in Example 4, a 1–1 correspondence between the points on AB and the points on $A'B'$, Then, as in Example 4 again, set up a 1–1 correspondence between the points $A'B'$ and the points of AC and apply

the proof of part (3) of Theorem 3.8.1. That is, use Figure 3.8.2 to pair any point P on AC with P' on $A'B'$ and then with P'' on AB.)

7. Given $S = \{a, b, c\}$, $T = \{1, 2, 3, 4\}$, $S' = \{k, l, m\}$, $T' = \{6, 7, 8, 9\}$; (a) write a 1–1 correspondence between S and S' and (b) one between T and T', and then use your 1–1 correspondences to (c) write a 1–1 correspondence between $S \cup T$ and $S' \cup T'$.

8. Given $S = \{1, 2\}$, $T = \{1, 3, 4\}$, $S' = \{2, 3\}$, $T' = \{2, 3, 4\}$, (a) write a 1–1 correspondence between S and S' and (b) one between T and T', and then use your 1–1 correspondences to (c) write a 1–1 correspondence between $S \times T$ and $S' \times T'$.

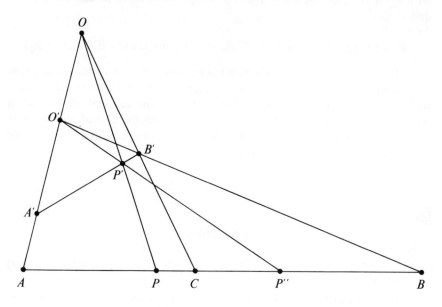

Figure 3.8.2

chapter 4

The whole numbers

4.1 Cardinal numbers, whole numbers, finite and infinite sets

We are now in a position to define the term **whole number** and to obtain some of the well-known basic facts about whole numbers.

If S is a given set, then for any set T, either S is equivalent to T or S is not equivalent to T. Some sets are equivalent to T and others are not. The set S has a certain property in common with those and only those sets that are equivalent to S. It is this property that we shall use to define what is mean by the **number of elements in** S.

Definition 4.1.1. *The property that a set S has in common with all of those sets and only those sets that are equivalent to S is called the **number of elements** in S or the **cardinal number** of S or the **number** of S.*

More rigorous definitions* (whose understanding requires a much more extensive mathematical background than this text assumes) of the term cardinal number are usually given in more advanced courses.

We will use the symbol $n(S)$ to denote the number of elements in a set S. In cases where the set is expressed in the form $\{a, b, \ldots, k\}$, we will also use $n\{a, b, \ldots, k\}$ to denote the number of elements in the set. It is important to realize that a set S is different from $n(S)$, the number of elements in S.

Now that we have defined what is meant by the number of elements in a set, let us consider what it should mean for sets to have the *same number of*

---

* See for example Alexander Abian, *The Theory of Sets and Transfinite Arithmetic*, W. B. Saunders Co., 1965.

elements. As we have illustrated in Section 3.8 and as Definition 4.1.1 indicates, we certainly want sets to have the same number of elements if the sets are equivalent. It would be quite natural to define sets to have the same number of elements (or the same cardinal number) if they are equivalent, and this is just what we do.

Definition 4.1.2. (*Definition of equality of cardinal numbers of sets*) *Let S and T be sets, then* $n(S) = n(T)$ *if and only if* $S \sim T$.

We will now define the familiar **whole numbers**. We begin by defining zero (written as 0) to be the number of elements in the null set, that is, $n(\emptyset) = 0$. We then define the number 1 to be $n\{0\}$. Of course, $\{0\}$ is not $\emptyset$ and since $\{0\}$ is not equivalent to $\emptyset$, $0 \neq 1$ by Definition 4.1.2. Now that we have 0 and 1, we define 2 to be $n\{0, 1\}$. We then use 0, 1, and 2 to define 3 as $n\{0, 1, 2\}$, and in a similar way we define 4 to be $n\{0, 1, 2, 3\}$, and we continue thus, defining 5, 6, 7, ... For each number h so defined, we define another number called the **successor** of h as $n\{0, 1, 2, \ldots, h\}$. Now, 1 is the successor of 0, 2 is the successor of 1, 3 is the successor of 2, etc. Each of these numbers is either 0 or the successor of a number. In each case the usual familiar symbol is used, for example $23 = n\{0, 1, 2, \ldots, 22\}$.

We formally summarize these definitions as follows:

Definition 4.1.3. *The whole numbers are the numbers* 0, 1, 2, 3, ..., *where*

$$0 = n(\emptyset); \qquad 3 = n\{0, 1, 2\};$$
$$1 = n\{0\}; \qquad 4 = n\{0, 1, 2, 3\};$$
$$2 = n\{0, 1\}; \qquad \text{etc.}$$

Definition 4.1.4. *The* **successor** *of any whole number h is the number*
$$n\{0, 1, 2, \ldots, h\}.$$

Every whole number therefore has a successor and is either zero or the successor of another whole number. We will use the symbol W to denote the set of whole numbers.

Now that we have the whole numbers defined, we notice that $n\{1\} = 1$; $n\{1, 2\} = 2$; $n\{1, 2, 3\} = 3$; $n\{1, 2, \ldots, s\} = s$; ..., because each of the sets can be put in 1–1 correspondence with the appropriate set used in Definition 4.1.3. For example, $\{0, 1, 2\} \sim \{1, 2, 3\}$ using the pairing $0 \leftrightarrow 1, 1 \leftrightarrow 2, 2 \leftrightarrow 3$, so $n\{1, 2, 3\} = n\{0, 1, 2\} = 3$.

The numbers 1, 2, 3, ... are called **counting numbers**. Sets of the form $\emptyset$ and $C_s = \{1, 2, 3, \ldots, s\}$ are called **counting sets**, because when we count the elements of a set H we do so by putting a set of the form $\{1. 2, 3, \ldots, s\}$ in 1–1 correspondence with H. For example, $n\{a, b, c, d\} = 4$, because we can count the elements of $\{a, b, c, d\}$ by putting it in 1–1 correspondence with $C_4 = \{1, 2, 3, 4\}$, with the pairing $1 \leftrightarrow a, 2 \leftrightarrow b, 3 \leftrightarrow c, 4 \leftrightarrow d$. The set $\{1, 2, 3, \ldots\}$ is not called a counting set, however, because it never ends.

Definition 4.1.5. *If the cardinal number of a set S is a whole number, then S is called a **finite set**; otherwise, S is called an **infinite set**.*

Thus a set is finite if and only if it can be put in 1–1 correspondence with a counting set. For example, the sets $\{a, b, c\}$ and $\{$Bill, George, Δ, ?$\}$ are finite sets and the sets $\{1, 2, 3, \ldots\}$ and $\{2, 4, 6, \ldots\}$ are infinite sets. The set of points on a line segment AB is also an infinite set. It is beyond the scope of this book to give a proof, but it can be proved that a set is infinite if and only if it can be put in 1–1 correspondence with a proper subset of itself, Therefore a set is finite if and only if it cannot be put in 1–1 correspondence with any proper subset of itself. For example, the infinite set $R = \{1, 2, 3, \ldots\}$ can be put in 1–1 correspondence with the proper subset $\{2, 4, 6, \ldots\}$ of itself, as we saw in Example 3 of Section 3.8.

Similarly, Exercise 6 of Section 3.8 demonstrates that the set of points on a line segment can be put in 1–1 correspondence with a proper subset of itself.

Although our major interest in this chapter is the system of whole numbers, it is of value to know a little more about cardinal numbers of infinite sets. These cardinal numbers are called **transfinite** cardinal numbers. In particular the cardinal number of the set $\{1, 2, 3, \ldots\}$ of nonzero whole numbers is the smallest transfinite cardinal number and is usually denoted by $\aleph_0$ (the Hebrew letter aleph with a subscript 0) and is read "aleph null." Of course any set equivalent to $\{1, 2, 3, \ldots\}$ also has the cardinal number $\aleph_0$. Sets of this kind are called **denumerable** or **enumerable**.

There are infinite sets which are not denumerable however. The set of points on a line segment AB is nondenumerable. The cardinal number of this set is often denoted by the letter c. The cardinal number c is larger than $\aleph_0$ in the sense that the set $\{1, 2, 3, \ldots\}$ can be put in 1–1 correspondence with a proper subset of the set of points on a line segment, but cannot be put in 1–1 correspondence with the whole set. In fact it can be proved that for any transfinite cardinal number, there is always a larger one.

Exercises

1. Write explicitly the definitions of the numbers 5, 6, 7, 8, and 9 and the counting sets C_5, C_6, C_7, C_8, and C_9.
2. Set up 1–1 correspondences to show that $n(C_5) = 5$, $n(C_6) = 6$, $n(C_7) = 7$, $n(C_8) = 8$, and $n(C_9) = 9$.
3. Prove that $n\{a, b, c, d, e, f, g\} = 7$.
4. Explain carefully how the following differ:
   (a) $\{0, 1, 2\}$; (b) $n\{0, 1, 2\}$; (c) $\{1, 2, 3\}$.
5. Are $n\{0, 1, 2\}$ and $n\{1, 2, 3\}$ different? Why?
6. Why is $3 \neq \{0, 1, 2\}$?

4.2   Addition and multiplication of whole numbers

Now that we have defined whole numbers, we can define addition and multiplication of whole numbers. Because our primary interest is in the whole
numbers, we will restrict our attention to the cardinal numbers of finite sets
in our definitions of addition and multiplication, although the same definitions
are used for infinite cardinal numbers.

Definition 4.2.1. (*Definition of addition in W.*) *Let a and b be any whole
numbers and let S and T be disjoint sets such that* $n(S) = a$ *and* $n(T) = b$; *then*

$$a + b = n(S \cup T).$$

Definition 4.2.2. (*Definition of multiplication in W.*) *Let a and b be any whole
numbers and let S and T be sets* (*not necessarily disjoint*) *such that* $n(S) = a$ *and*
$n(T) = b$; *then*

$$a \times b = n(S \times T).$$

We call $a + b$ the **sum** of a and b, while the individual numbers a and b are
called the **summands** or **addends** or **terms** of $a + b$. The expression $a \times b$ is often
written in the form $a \cdot b$ or just ab and is called the product of a and b, and here
the individual numbers a and b are called the **factors** of ab.

We are now able to see why numbers are added and multiplied in the
familiar way. For example, we can see why $2 + 3 = 5$. We must first obtain
disjoint sets S and T such that $n(S) = 2$ and $n(T) = 3$. Let us take $S = \{1, 2\}$
and $T = \{a, b, c\}$, then $S \cap T = \varnothing$. Also $n(S) = 2$ because S is the counting
set C_2 and $n(T) = 3$ because $T \sim C_3$. Now

$$2 + 3 = n(S) + n(T) = n(S \cup T) = n\{1, 2, a, b, c\}.$$

But $\{1, 2, a, b, c\} \sim \{1, 2, 3, 4, 5\}$, so it has 5 elements and hence,

$$2 + 3 = n\{1, 2, a, b, c\} = 5,$$

and thus we have seen why $2 + 3 = 5$; in fact we have proved that $2 + 3 = 5$.

Let us see what would happen if we did not take disjoint sets. Suppose
that we take $S = C_2 = \{1, 2\}$ and $T = C_3 = \{1, 2, 3\}$. Then $S \cap T = \{1, 2\} \neq \varnothing$
and in this case $n(S \cup T) = n(\{1, 2\} \cup \{1, 2, 3\}) = n\{1, 2, 3\} = 3 \neq 5$.

We can, however, use these sets C_2 and C_3 to see why $2 \times 3 = 6$, because
for multiplication it is not necessary to use disjoint sets. Let $S = C_2$ and $T = C_3$

as before. Then $n(S) = 2$ and $n(T) = 3$ and $2 \times 3 = n(S) \times n(T) = n(S \times T) =$ $n\{(1, 1), (1, 2), (1, 3), (2, 1), (2, 2), (2, 3)\} = 6$, because $\{(1, 1), (1, 2), (1, 3),$ $(2, 1), (2, 2), (2, 3)\}$ is equivalent to the counting set $C_6 = \{1, 2, 3, 4, 5, 6\}$. We have thus seen why $2 \times 3 = 6$ and have also proved that $2 \times 3 = 6$.

In the same way, we can prove that all of the familiar sums and products are in fact equal to the appropriate numbers that we have memorized them to be.

Notice that it follows from Definition 4.2.1 that if S and T are any finite sets such that $S \cap T = \varnothing$, then

$$n(S \cup T) = n(S) + n(T).$$

The reason for this is as follows: Since S and T are finite sets, $n(S)$ and $n(T)$ are in W. Let $n(S) = a$ and $n(T) = b$, then $a + b = n(S) + n(T) = n(S \cup T)$.

Suppose that a and b are whole numbers. Then there are many sets with cardinal number a and many sets with cardinal number b. It might at first appear that the definitions of $a + b$ and ab depend on the particular sets S and T and may not produce a unique sum or product. It might be the case that if different sets S' and T' such that $n(S') = a$ and $n(T') = b$ are used, a different $a + b$ or a different ab is obtained. In one case, $a + b$ is $n(S \cup T)$ and $ab = n(S \times T)$ while in the other case $a + b$ is $n(S' \cup T')$ and ab is $n(S' \times T')$. The definitions of $a + b$ and ab might thus seem ambiguous. They are not in fact ambiguous, however, because we can prove the following theorems:

Theorem 4.2.1. *If S and T are disjoint sets and S' and T' are disjoint sets such that $S \sim S'$ and $T \sim T'$, then $n(S \cup T) = n(S' \cup T')$.*

Proof. This theorem follows directly from Theorem 3.8.3, because by that theorem if $S \sim S'$ and $T \sim T'$, then $S \cup T \sim S' \cup T'$. Hence, by Definition 4.1.2 we have $n(S \cup T) = n(S' \cup T')$.

Theorem 4.2.2. *If S and T are sets (not necessarily disjoint) and S' and T' are sets such that $S \sim S'$ and $T \sim T'$, then $n(S \times T) = n(S' \times T')$.*

Proof. Similarly to the proof of Theorem 4.2.1, it follows from Theorem 3.8.3 that if $S \sim S'$ and $T \sim T'$, then $S \times T \sim S' \times T'$. Hence $n(S \times T) = n(S' \times T')$ by Definition 4.1.2.

Addition and multiplication of whole numbers are therefore independent of the particular sets S and T that are used in the definitions of addition and multiplication. Theorems 4.2.1 and 4.2.2 can be stated as follows: If a, b, a', and b' are whole numbers such that $a = a'$ and $b = b'$, then $a + b = a' + b'$ and $ab = a'b'$. This situation is described by saying that addition and multiplication of whole numbers are **well defined.**

Exercises

Show why each of the following is true:

1. $1 + 1 = 2$;   **2.** $2 + 1 = 3$;   **3.** $3 + 4 = 7$;

4. $3 + 0 = 3$;   **5.** $2 \times 2 = 4$;   **6.** $3 \times 4 = 12$;   **7.** $4 \times 0 = 0$.

8. If $S = \{1, 2, 3, 4, 5\}$ and $T = \{2, 4, 7, 8\}$, find $n(S \cup T)$ and $n(S \cap T)$ and verify that $n(S \cup T) + n(S \cap T) = n(S) + n(T)$. (Actually, this is true for all finite sets S and T).

4.3   Properties of addition and multiplication

We shall now see why addition and multiplication of whole numbers have certain familiar properties. Although the names of these properties may not be familiar to the reader, they are in fact called the **closure**, **commutative**, **associative**, **identity**, and **distributive** properties.

Before proceeding directly with these properties, let us first take note of something that will help us to see why they hold. As a specific example, notice that the counting set $C_4 = \{1, 2, 3, 4\}$ is equivalent to $C_4 \times \{1\} = \{(1, 1), (2, 1), (3, 1), (4, 1)\}$ by virtue of the 1–1 correspondence $1 \leftrightarrow (1, 1)$, $2 \leftrightarrow (2, 1)$, $3 \leftrightarrow (3, 1)$, $4 \leftrightarrow (4, 1)$. Also notice that the intersection of $C_4 \times \{1\}$ with every counting set is $\varnothing$, because the elements of $C_4 \times \{1\}$ are ordered pairs of whole numbers while the elements in each counting set except $\varnothing$ are whole numbers.

More generally, if we take any counting set $C_s = \{1, 2, \ldots, s\}$, it is equivalent to $C_s \times \{1\} = \{(1, 1), (2, 1), \ldots, (s, 1)\}$ because of the 1–1 correspondence $1 \leftrightarrow (1, 1)$, $2 \leftrightarrow (2, 1)$, $\ldots$, $s \leftrightarrow (s, 1)$. Also the intersection of $C_s \times \{1\}$ with every counting set is $\varnothing$ because an ordered pair of whole numbers is not a whole number.

We now proceed with the properties of addition and multiplication.

Definition 4.3.1. *A whole number that is not zero is called a **positive whole number** or a **positive integer** or a **natural number** or a **counting number**.*

From now on we will use P to denote the set $\{1, 2, \ldots\}$ of all positive whole numbers. We will now show that P **is closed under addition**, that is, that *the sum $a + b$ of any positive whole numbers a and b is always a positive whole number*. Although we will also state as part of the next theorem that P **is closed under multiplication** (that is, that the *product ab of any positive whole numbers a and b is always a positive whole number*), it is convenient to defer the proof until the next section.

Theorem 4.3.1. *P is closed under addition and multiplication.*

Proof. As mentioned above, we will only prove here that P is closed under addition.

Let a and b be any positive whole numbers, and consider the counting sets $C_a = \{1, 2, \ldots, a\}$ and $C_b = \{1, 2, \ldots, b\}$. (Here, of course, C_a is the set of whole numbers from 1 to a and C_b is the set of whole numbers from 1 to b). We cannot use both C_a and C_b to get $a + b$, because $C_a \cap C_b \neq \varnothing$. We have seen above that $C_b \sim C_b \times \{1\}$ and that $C_a \cap (C_b \times \{1\}) = \varnothing$, so we will use C_a and $C_b \times \{1\}$ to get $a + b$. Now $a + b = n(C_a) + n(C_b) = n(C_a) + n(C_b \times \{1\}) = n(C_a \cup (C_b \times \{1\})) = n\{1, 2, \ldots, a, (1, 1), (2, 1), \ldots, (b. 1)\}$. In order to see that $a + b \in P$, we need to see that the set $H = \{1, 2, \ldots, a, (1, 1), (2, 1), \ldots, (b, 1)\}$ can be put in 1–1 correspondence with some non-empty counting set. To see this, list the counting sets in the following natural order, that is, the order in which they were defined.

$$C_1 = \{1\},$$
$$C_2 = \{1, 2\},$$
$$C_3 = \{1, 2, 3\},$$
$$C_4 = \{1, 2, 3, 4\},$$
$$C_5 = \{1, 2, 3, 4, 5\}, \quad \text{etc.}$$

We will eventually get to $C_a = \{1, 2, \ldots, a\}$.

Now the first counting set after C_a on this list can be put in 1–1 correspondence with the set $\{1, 2, \ldots, a, (1, 1)\}$. The second one after C_a on this list can be put in 1–1 correspondence with $\{1, 2, \ldots, a, (1, 1), (2, 1)\}$, etc.; the bth counting set after C_a on this list can be put in 1–1 correspondence with

$$\{1, 2, \ldots, a, (1, 1), (2, 1), \ldots, (b, 1)\} = H$$

Therefore, $a + b$ is the positive whole number that is the number of elements in the bth counting set after the set C_a in the above listing. Therefore, P is closed under addition.

Theorem 4.3.2. *If $a, b, c \in W$, then*
(1) $a + b \in W$ (*W is closed under addition*);
(2) $a + b = b + a$ (*Addition of whole numbers in commutative*);
(3) $(a + b) + c = a + (b + c)$ (*Addition in W is associative*);
(4) $a + 0 = 0 + a = a$ (*Zero is the **identity of addition***);
(5) $ab \in W$ (*W is closed under multiplication*);
(6) $ab = ba$ (*Multiplication in W is commutative*);
(7) $(ab)c = a(bc)$ (*Multiplication in W is associative*);
(8) $a \cdot 1 = 1 \cdot a = a$ (*1 is the **identity of multiplication***);
(9) $a(b + c) = ab + ac$, and $(b + c)a = ba + ca$ (*Multiplication in W is distributive on both sides over addition*).

Proof. Throughout this proof we will let R, S, and T be disjoint sets such that $n(R) = a$, $n(S) = b$, and $n(T) = c$. Such sets exist; in fact, we could take $R = C_a$, $S = C_b \times \{1\}$, and $T = C_c \times \{2\}$.

Proof of (2). $a + b = b + a$.

Using the definition of addition and the fact that $R \cup S = S \cup R$ we have $a + b = n(R \cup S) = n(S \cup R) = b + a$ and, hence, $a + b = b + a$.

Proof of (3). $(a + b) + c = a + (b + c)$.

Again by the definition of addition and the fact that $(R \cup S) \cup T = R \cup (S \cup T)$, we get $(a + b) + c = n(R \cup S) + n(T) = n((R \cup S) \cup T) = n(R \cup (S \cup T)) = n(R) + n(S \cup T) = a + (b + c)$. Therefore, $(a + b) + c = a + (b + c)$.

Proof of (4). $a + 0 = 0 + a = a$.

By part (2), $a + 0 = 0 + a$. Also, by the definition of addition and the fact that $R \cup \varnothing = R$, we have $a + 0 = n(R \cup \varnothing) = n(R) = a$, and hence $a + 0 = 0 + a = a$.

Proof of (1). $a + b \in W$.

Either both a and b are different from 0 or at least one of them is zero. In the first case a and b are both in P, so $a + b \in P$ by Theorem 4.3.1. But $P \subset W$ so $a + b \in W$. In the second case either $a = 0$ or $b = 0$. If $a = 0$, then $a + b = 0 + b = b$ by part (4), so $a + b = b \in W$. If $b = 0$, then $a + b = a + 0 = a$ by part (4), so $a + b = a \in W$. Therefore, in all cases $a + b \in W$.

Proof of (6). $ab = ba$.

By Theorem 3.8.2, $R \times S \sim S \times R$, so $n(R \times S) = n(S \times R)$. Therefore, $ab = n(R \times S) = n(S \times R) = ba$, and we have $ab = ba$.

Proof of (7). $(ab)c = a(bc)$.

By Theorem 3.8.2, $(R \times S) \times T \sim R \times (S \times T)$, so $n((R \times S) \times T) = n(R \times (S \times T))$. Therefore, $(ab)c = n((R \times S) \times T) = n(R \times (S \times T)) = a(bc)$, and we have $(ab)c = a(bc)$.

Proof of (8). $a \cdot 1 = 1 \cdot a = a$.

By part (6), $a \cdot 1 = 1 \cdot a$. Also, $a \cdot 1 = n(R \times \{1\})$. Now, if we pair each $s \in R$ with $(s, 1)$ of $R \times \{1\}$, we get a 1–1 correspondence between R and $R \times \{1\}$, so $R \sim R \times \{1\}$. Hence, we have $n(R \times \{1\}) = n(R) = a$. Therefore, $a \cdot 1 = 1 \cdot a = a$.

Proof of (9). $a(b + c) = ab + ac$ and $(b + c)a = ba + ca$.

By Theorem 3.7.2, $R \times (S \cup T) = (R \times S) \cup (R \times T)$, so $a(b + c) = n(R \times (S \cup T)) = n((R \times S) \cup (R \times T)) = n(R \times S) + n(R \times T) = ab + ac$.

Therefore, $a(b + c) = ab + ac$. To prove that $(b + c)a = ba + ca$, we use part (6) twice and part (9) to get $(b + c)a = a(b + c) = ab + ac = ba + ca$.

We will leave to Section 4.4 the proof of part (5) that $ab \in W$.

The above fundamental properties can be used to prove relations and generalizations such as those of Section 4.4. In order to obtain a better understanding of the commutative, associative, and distributive properties and their use in establishing other relations, the reader should use them to prove a few other relations such as those in the exercises at the end of this section. By way of illustration, we shall now prove that if p, q, and r are in W, then $(rp)q = (qp)r$.

Proof. By the associative property of multiplication

$$(rp)q = r(pq),$$

then using the commutative property of multiplication we have

$$r(pq) = (pq)r.$$

Again using the commutative property of multiplication

$$(pq)r = (qp)r.$$

Hence we have shown that

$$(rp)q = r(pq) = (pq)r = (qp)r,$$

and therefore

$$(rp)q = (qp)r.$$

Exercises

1. Decide whether each of the following sets is closed under addition.
   (a) The set of even whole numbers;
   (b) The set of odd whole numbers;
   (c) The set of whole numbers each of which is a multiple of 3;
   (d) $\{5, 10, 15, 20, 25\}$;
   (e) $\{5, 10, 15, 20, 25, \ldots\}$;
   (f) $\{x \in W \mid x > 10\}$.
2. Decide whether each of the sets in Exercise 1 is closed under multiplication.

3. Use the commutative, associative, and distributive properties to prove that if
$p, q, r, s, \in W$, then

(a) $(p + q) + r = q + (p + r)$;

(b) $p((q + r) + s) = pq + (pr + ps)$;

(c) $(p + q) + (r + s) = (p + s) + (r + q)$;

(d) $s((pq)r) = (qs)(rp)$.

4. Show that if $h \in W$ and $h + p = p$ for all $p \in W$, then $h = 0$. (*Hint*; Use part (4)
of Theorem 4.3.2 and the fact that if $h + p = p$ for all $p \in W$, then $h + 0 = 0$.)

5. Show that if $h \in W$ and $hp = p$ for all $p \in W$, then $h = 1$. (*Hint*: Use part (8)
of Theorem 4.3.2.)

4.4   More general associative, commutative, and distributive properties

The operations of union and intersection of sets and addition and multiplication of whole numbers are methods of obtaining a third thing from **two** things,
that is, they are **binary operations**. We must know to which two things the
operation applies in each case. Parentheses are used where necessary to make
it clear which things are combined. For example, $(R \cup S) \cap T$ means something different from $R \cup (S \cap T)$. In some cases, however, the parentheses can be eliminated. For example, since for whole numbers, $a + (b + c) =
(a + b) + c$, it does not matter which two things are combined first, and we
can write $a + b + c$ without parentheses for either or both of them. Similarly,
several applications of the associative property of addition will show that
$(a + b) + (c + d) = ((a + (b + c)) + d = a + (b + (c + d))$ and in fact the same
sum is obtained no matter how the numbers are associated. Hence, there is no
ambiguity in writing $a + b + c + d$. The same is true for any *finite* sum or
product of whole numbers or union or intersection of sets. It is the associative
property that makes this possible. For example, we can unambiguously write
any of the following, where the S's are sets and the a's are whole numbers:

(1) $S_1 \cup S_2 \cup S_3 \cup S_4 \cup S_5$.

(2) $S_1 \cap S_2 \cap S_3 \cap S_4$.

(3) $a_1 + a_2 + a_3 + a_4 + a_5 + a_6$.

(4) $a_1 a_2 a_3 a_4 a_5$.

By using the commutative property, we do not even need to write these in
any particular order. For example, (3) above could also be written as

$$a_3 + a_2 + a_5 + a_6 + a_1 + a_4.$$

We must be careful, however, if our expression involves more than one
operation, because we cannot usually remove all parentheses unless it is

agreed in advance that some particular unambiguous way of combining the terms is always used.

It is common to make such an agreement for combinations of addition and multiplication. The accepted agreement is to perform all multiplications before the additions, and we will accept this agreement. For example,

$$3 + 4 \times 7 + 6 \times 2 \text{ means } 3 + (4 \times 7) + (6 \times 2) = 3 + 28 + 12 = 43.$$

Another property that can be generalized is the distributive property. Now that we have an unambiguous meaning for any finite sum of the form

$$b_1 + b_2 + b_3 + \ldots + b_h$$

where h is a positive whole number (the number of b's being added) and the b's are any whole numbers, we can repeatedly use the distributive property to obtain

$$a(b_1 + b_2 + \ldots + b_h) = ab_1 + ab_2 + \ldots + ab_h$$

for any whole number a.

It should also be pointed out that it follows from the definitions of successor and addition that the successor of a whole number is that number plus 1. For example, the successor of 7 (which is called 8) is $7 + 1$.

From the definition of addition it also follows that $1 + 1 + 1 + 1 = 4$, and that in general the sum of s numbers each equal to 1 is s. This latter fact can be used to see that $2 \times 3 = 2(1 + 1 + 1) = 2 + 2 + 2$ and $2 \times 3 = 3 \times 2 = 3(1 + 1) = 3 + 3$, so that 2×3 is equal to the sum of three 2's or the sum of two 3's.

Similarly, for any two positive whole numbers a and b (using the notation $(d + d + \ldots + d)_h$ to mean the sum of h numbers each equal to d)

$$a \cdot b = a(1 + 1 + \ldots + 1)_b = (a + a + \ldots + a)_b$$
$$a \cdot b = b \cdot a = b(1 + 1 + \ldots + 1)_a = (b + b + \ldots + b)_a.$$

Thus $a \times b$ is equal to the sum of a numbers each equal to b, or the sum of b numbers each equal to a.

In particular, $a + a = 2a$ and $b + b + b = 3b$. Because, by part (1) of Theorem 4.3.2, the sum of whole numbers is a whole number, we now know that the product of two whole numbers, neither of which is zero, is a whole number. In fact, by the addition part of Theorem 4.3.1 it is a positive whole number, and we have completed the proof of Theorem 4.3.1 that the product of two numbers of P is in P.

Since a whole number is in P if and only if it is not zero, the multiplication part of Theorem 4.3.1 is equivalent to the statement that for a and b in W, if $a \neq 0$ and $b \neq 0$ then $ab \neq 0$. Now this implication is equivalent to its contrapositive, that is, it is equivalent to the statement that if $ab = 0$ then $a = 0$ or $b = 0$.

What about the converse of this last implication? That is, must a product equal zero if one of the factors is zero? The next theorem assures us that it must be. This result, along with the fact that P is closed under multiplication, completes the proof of part (5) of Theorem 4.3.2 that W is closed under multiplication.

Theorem 4.4.1. *For a and b in W, $ab = 0$ if and only if $a = 0$ or $b = 0$.*

Proof. We have noted above that the implication—if $ab = 0$ then $a = 0$ or $b = 0$—is equivalent to part of Theorem 4.3.1 and hence we only need to prove that if $a = 0$ or $b = 0$, then $ab = 0$. First let $b = 0$ and let S be a set such that $n(S) = a$. Of course $n(\emptyset) = 0$ by the definition of 0, and thus $a \cdot 0 = n(S \times \emptyset) = n(\emptyset) = 0$, because $S \times \emptyset = \emptyset$ by Theorem 3.7.1. Now if $a = 0$, let T be a set such that $n(T) = b$, and then we similarly have $0 \cdot b = n(\emptyset \times T) = n(\emptyset) = 0$, and we have completed the proof.

This theorem is not only useful in the completion of the proof of Theorem 4.3.2, part (5), but it is especially interesting in its own right. It is important to realize that this is not just a rule that someone dictated, but it is something that we can prove.

Exercises

1. Compute: (a) $3 \times 4 + 7 \times 8 + 2$;
   (b) $2 + 5 \times 3 \times 2 + 7 \times 2$.
2. Use the distributive property (more than once) to prove that, for any numbers $a, b, c, d \in W$, $(a + b)(c + d) = ac + ad + bc + bd$.
3. Use the distributive property (more than once) to prove that, for $a, b, \in W$, $(a + b)(a + b) = aa + 2ab + bb$.

4.5   The number line

As a form of picture for the whole numbers, we will put them in a 1–1 correspondence with a certain subset of the set of points on a line.

Let us take a line, select some point on it, pair it with 0, and label it 0. Then with some convenient unit of measure, pair the point 1 unit to the right of the 0 point with 1, the point 2 units to the right of the 0 point with 2, ..., the point h units to the right of the 0 point with $h, \ldots$. Some of these labeled points are shown in Figure 4.5.1. We thus have a 1–1 correspondence between W and a subset of the points on the line. When a point is labeled h in the fashion described, we will often call it h or the **point** h. The whole labeled line will be called the **number line**. The point 0 on the number line will also be called the **origin**. In later chapters we shall eventually label all the points on the line with numbers as we obtain larger and larger systems of numbers.

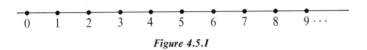

Figure 4.5.1

Since any point h is h units to the right of the origin, we note that $a + b = c$ if and only if c is b units to the right of a and a is b units to the left of c.

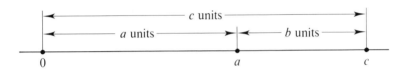

Thus we could find $a + b$ on the number line mechanically by going b units to the right of a.

Similarly, using the fact mentioned in Section 4.4 that ab is equal to the sum of a numbers each equal to b, we could find $a \times b$ on the number line mechanically by going a steps of b units each to the right of the origin.

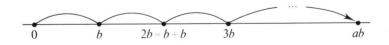

For example, to find $3 + 2$ on the number line, go 2 units to the right of 3.

We arrive at 5, the correct result. To find 3 × 2 on the number line, go 3 steps of 2 units each to the right of the origin.

We arrive at the result 6.

Exercises

1. On the number line, how many units to the right of 86 is 113?

2. Draw figures showing how you can find each of the following on the number line mechanically: (a) 4 + 3; (b) 4 × 3; (c) 5 + 6; (d) 5 × 6; (e) 0 + a; (f) a + a + a; (g) a × 4; (h) 4a; (i) cd; (j) c + d.

4.6   Exponents

When one has a product $a \cdot a \cdot \ldots \cdot a$ of h numbers each equal to a, it is a great convenience to express this product as a^h. This is easier to read and write and also the h tells how many a's are multiplied in the product.

Definition 4.6.1. *For any $a \in W$ and any positive whole number h, a^h means the product of h numbers each equal to a.*

It is a convenience to define a^h when $a \neq 0$ and $h = 0$ as follows:

Definition 4.6.2. *For any positive whole number a, $a^0 = 1$.*

There is no useful way to define 0^0, so it is left as a meaningless combination of symbols.

Examples. (1) $3^2 = 3 \cdot 3 = 9$; (2) $2^3 = 2 \cdot 2 \cdot 2 = 8$; (3) $2^1 = 2$; (4) $7^1 = 7$; (5) $4^0 = 1$; (6) $7^0 = 1$; (7) $2^4 = 2 \cdot 2 \cdot 2 \cdot 2 = 16$; (8) $0^1 = 0$; (9) $3^1 = 3$; (10) $1^0 = 1$; (11) 0^0 has no meaning; (12) $0^3 = 0 \cdot 0 \cdot 0 = 0$.

In the notation a^h, h is called the **exponent** and a is called the **base**. The expression a^h is read as "a to the hth power" or as "a to the hth" or as "a exponent h." Thus 3^{17} is read "3 to the 17th power" or "3 to the 17th" or "3 exponent 17."

When the exponent is 2 or 3, however, it is usually read differently. Although a^2 can be read as "a to the second power," it is usually read as "a square," and a^3 is usually read "a cube." These are related to squares and cubes, because a square 6 feet on a side has an area of 6^2 square feet and a cube 4 inches on a side has a volume of 4^3 cubic inches.

It is necessary to know how to simplify expressions involving exponents. For example.

$$(1)\ 3^2 \cdot 3^3 = (3 \cdot 3)(3 \cdot 3 \cdot 3) = 3 \cdot 3 \cdot 3 \cdot 3 \cdot 3 = 3^5.$$
$$(2)\ (5^2)^3 = 5^2 \cdot 5^2 \cdot 5^2 = (5 \cdot 5)(5 \cdot 5)(5 \cdot 5) = 5 \cdot 5 \cdot 5 \cdot 5 \cdot 5 \cdot 5 = 5^6$$
$$(3)\ (2 \cdot 5)^3 = (2 \cdot 5)(2 \cdot 5)(2 \cdot 5) = 2 \cdot 2 \cdot 2 \cdot 5 \cdot 5 \cdot 5 = 2^3 \cdot 5^3.$$

More generally, we have the following theorem.

Theorem 4.6.1. *If a, b, r, $s \in W$, then*
(1) $a^r \cdot a^s = a^{r+s}$ *(the exponents are added)*,
(2) $(a^r)^s = a^{rs}$ *(the exponents are multiplied)*,
(3) $(ab)^r = a^r b^r$,

(except that if 0^0 occurs in any one of these, that relationship becomes meaningless).

Proof. We assume that none of the exponents are 0 and leave as exercises the cases where one or more of them is zero. We use the notation $(a \cdot a \cdot \ldots \cdot a)_r$ to denote the product of r factors each equal to a.

Proof of (1). $a^r \cdot a^s = a^{r+s}$.
$$a^r \cdot a^s = (a \cdot a \cdot \ldots \cdot a)_r (a \cdot a \cdot \ldots \cdot a)_s = (a \cdot a \cdot \ldots \cdot a)_{r+s} = a^{r+s}.$$

Proof of (2). $(a^r)^s = a^{rs}$.
$$(a^r)^s = (a^r \cdot a^r \cdot \ldots \cdot a^r)_s = [(a \cdot a \cdot \ldots \cdot a)_r (a \cdot a \cdot \ldots \cdot a)_r$$
$$\ldots (a \cdot a \cdot \ldots \cdot a)_r]_s = (a \cdot a \cdot \ldots \cdot a)_{rs} = a^{rs}$$

Proof of (3). $(ab)^r = a^r b^r$.
$$(ab)^r = [(ab)(ab) \ldots (ab)]_r = (a \cdot a \cdot \ldots \cdot a)_r (b \cdot b \cdot \ldots \cdot b)_r = a^r b^r.$$

Examples. (1) $3^2 \cdot 3^3 = 3^{2+3} = 3^5$; (2) $a^5 \cdot a^2 = a^{5+2} = a^7$; (3) $(5^2)^3 = 5^{2 \cdot 3} = 5^6$; (4) $(h^2)^4 = h^{2 \cdot 4} = h^8$; (5) $(2 \cdot 5)^3 = 2^3 \cdot 5^3$; (6) $(a^3 b^2)^3 = a^9 b^6$.

Exercises

1. Find (a) 3^4; (b) $(2^3)^2$; (c) $(a^4 b^6)^7$; (d) $(3^7)^0$; (e) $a^5 \cdot a^8$; (f) $(a^7)^5$; (g) $2^0 \cdot 2^3$; (h) $(a^3 b^0 c^5 d^6)^4$.

2. Find (a) 2^5; (b) $(3^2)^4$; (c) $h^7 \cdot h^9$; (d) $r^h \cdot r^k$; (e) $(d^{11})^8$; (f) $(x^2 y^3 z)^7$.

3. Prove Theorem 4.6.1 for the case where $r = 0$ and the case where $s = 0$, that is, prove that when $a \neq 0$ and $b \neq 0$, (a) $a^0 \cdot a^s = a^{0+s}$; (b) $(a^0)^s = a^{0 \cdot s}$; (c) $(ab)^0 = a^0 b^0$; (d) $a^r \cdot a^0 = a^{r+0}$; (e) $(a^r)^0 = a^{r \cdot 0}$.

4.7 Inequalities

We will now be concerned with the meaning of "*a* is less than *b*" for whole numbers *a* and *b*. We shall write "*a* is less than *b*" as $a < b$. It will have the same meaning as "*b* is greater than *a*" which we shall write as $b > a$.

Definition 4.7.1. (*Definition of "less than" for whole numbers.*) *If a and b are whole numbers, $a < b$ if and only if there exists a positive whole number p such that $a + p = b$.*

Definition 4.7.2. (*Definition of "greater than" for whole numbers.*) *If a and b are whole numbers, $b > a$ means the same thing as $a < b$.*

For example, $3 < 5$ (and $5 > 3$) because there exists a positive whole number *p* such that $3 + p = 5$. Also, if *p* is positive, then $p > 0$, and conversely, because $p = 0 + p$. So the positive whole numbers are those that are greater than zero.

Let us consider the significance of $a < b$ on the number line. Similarly to the discussion in Section 4.5, $a + p = b$ for a positive whole number *p* if and only if *a* is *p* units, where $p \neq 0$, to the left of *b* on the number line.

That is $a < b$ if and only if *a* is to the *left* of *b* on the number line. In a similar manner, since $a > b$ means $b < a$, $a > b$ if and only if *a* is to the *right* of *b* on the number line.

Returning to the general situation again where *a* and *b* are any particular whole numbers, let us consider their counting sets C_a and C_b. Notice that a unique one of the three relations $C_a = C_b$, C_a is a proper subset of C_b, or C_b is a proper subset of C_a holds. Now,

    (1) $C_a = C_b$ if and only if $a = b$.
    (2) C_a is a proper subset of C_b if and only if $a < b$.
    (3) C_b is a proper subset of C_a if and only if $b < a$.

The first of these three statements follows from the meaning of C_a while the second and third are less immediate. The third statement is proved similarly to the second so we will demonstrate only the second. We must show (i) that if C_a is a proper subset of C_b then $a < b$, and we must also show (ii) that if $a < b$ then C_a is a proper subset of C_b.

Proof of (i). If C_a is a proper subset of C_b, then $C_b = C_a \cup C$ where $C = C_b - C_a$ (recall that by Definition 3.4.1 this is the set of elements of C_b that are not in C_a). Here $C_a \cap C = \varnothing$ so

$$b = n(C_b) = n(C_a \cup C) = n(C_a) + n(C).$$

Now $n(C_a) = a$ and since $C \neq \varnothing$, $n(C) = a$ positive whole number p. Therefore, $b = a + p$ where p is positive, so $a < b$.

(As an illustration of this proof, take $C_3 = \{1, 2, 3\}$ and $C_5 = \{1, 2, 3, 4, 5\}$. Here C_3 is a proper subset of C_5 and $C_5 = C_3 \cup C$ where $C = C_5 - C_3 = \{4, 5\}$, and we have $5 = n(C_5) = n(C_3 \cup C) = n(C_3) + n(C) = 3 + 2$.)

Proof of (ii). If $a < b$ then there exists a positive whole number p such that $b = a + p$. Therefore,

$$C_b = C_a \cup \{a + 1, \ldots, a + p\},$$

and so C_a is a proper subset of C_b.

(As an illustration, $4 < 7$, $7 = 4 + 3$, and $C_7 = C_4 \cup \{5, 6, 7\}$.)

We have seen that if a and b are any particular whole numbers, then a unique one of the following must hold: $C_a = C_b$, C_a is a proper subset of C_b, C_b is a proper subset of C_a. However, by the statements (1), (2), and (3), $C_a = C_b$ is equivalent to $a = b$, C_a is a proper subset of C_b is equivalent to $a < b$, and C_b is a proper subset of C_a is equivalent to $b < a$, so we have proved the following theorem.

Theorem 4.7.1. *If a and b are whole numbers, then a unique one of the following must hold:*

    (i) $a = b$;   (ii) $a < b$;   (iii) $a > b$.

This theorem is known as the **trichotomy** property of ordering. Note that since a unique one of the relations must hold we know that at least one of them holds and at most one of them holds, For example, if $a < b$, then $a \neq b$ and $a \not> b$. In our number line picture, we can see that the trichotomy property is simply that a unique one of the following must hold: a and b are the same point; a is to the left of b; a is to the right of b.

We may wonder if the relation "$<$" is reflexive, symmetric, or transitive. It is not reflexive, because $a = a$ and so, by the trichotomy property, $a \not< a$. It is not symmetric because if $a < b$ then, by the trichotomy property again, $b \not< a$. It is transitive, however, and we have the following theorem.

Theorem 4.7.2. *If a, b, $c \in W$ and $a < b$ and $b < c$, then $a < c$. (The transitive property of $<$.)*

The number line interpretation of this theorem is: If a is to the left of b and b is to the left of c, then a is to the left of c.

Proof. If $a < b$ and $b < c$, then by the definition of $<$, there are positive whole numbers q and r such that $a + q = b$ and $b + r = c$. Therefore,

$$b + r = a + q + r = c.$$

But by Theorem 4.3.1 the sum $q + r$ of the positive whole numbers q and r is a positive whole number p such that $a + p = c$. Therefore, $a < c$.

Let us consider the following numerical example of our next theorem. We know that $2 < 5$ (because $2 + 3 = 5$). What happens if we add the same number, say 4, to each of the numbers 2 and 5? We get $2 + 4$ and $5 + 4$, and of course $2 + 4 < 5 + 4$, so the inequality is still in the same "direction." Similarly, if we multiply each of the numbers 2 and 5 by the same positive whole number, say by 3, we get $2 \cdot 3 = 6$ and $5 \cdot 3 = 15$ and $2 \cdot 3 < 5 \cdot 3$, so again the inequality keeps the same "direction."

The next theorem states that the situation we just considered is quite general.

Theorem 4.7.3. *If $a, b \in W$, $a < b$, and h is any whole number, then $a + h < b + h$.*

Proof. If $a < b$, then there exists a positive whole number p such that $a + p = b$. But then $a + p + h = b + h$, so $(a + h) + p = b + h$. Therefore, $a + h < b + h$ by the definition of "$<$", and we have proved the theorem.

Theorem 4.7.4. *If $a < b$ and h is a positive whole number, then $ah < bh$.*

Proof. Since $a + p = b$ for a positive p, $(a + p)h = bh$. But $(a + p)h = ah + ph$ by the distributive property, so we have $ah + ph = bh$. Now since the product of the two positive whole numbers p and h is positive, ph is positive. Therefore, by the definition of "$<$", $ah < bh$.

Theorem 4.7.5. *Assume $a, b, c \in W$. If $a + c = b + c$, then $a = b$.*

Proof. We will prove the contrapositive, namely, if $a \neq b$, then $a + c \neq b + c$. If $a \neq b$, then by the trichotomy property we must have either $a < b$ or $b < a$. If $a < b$, then, by Theorem 4.7.3, $a + c < b + c$ and hence $a + c \neq b + c$. If $b < a$, then, by Theorem 4.7.3 again, $b + c < a + c$ and hence $b + c \neq a + c$, which completes the proof.

The property expressed in Theorem 4.7.5 is called the **cancellation property of addition**. (It allows the c to be cancelled from both sides of the equation.) We will now obtain the **cancellation property of multiplication**.

Theorem 4.7.6. *Assume $a, b, h \in W$. If $ah = bh$, then $h = 0$ or $a = b$.*

Proof. As in Theorem 4.7.5, we will prove the contrapositive, namely, if $a \neq b$ and $h \neq 0$, then $ah \neq bh$. Of course, in W, to say that $h \neq 0$ is to say that h is positive. Now if $a \neq b$, then by the trichotomy property we must have either $a < b$ or $b < a$. In the case where $a < b$, then, by Theorem 4.7.4, $ah < bh$ so $ah \neq bh$. Similarly, if $b < a$, then $bh < ah$ so $bh \neq ah$. We have now proved that if $a \neq b$ and $h \neq 0$, then $ah \neq bh$. Therefore, we have shown that if $ah = bh$, then $h = 0$ or $a = b$.

Notice that Theorem 4.7.6 can be restated as follows: *If $ah = bh$ and $h \neq 0$, then $a = b$.*

Example 1. If $x + 5 = 7$, then (since $7 = 2 + 5$) we have $x + 5 = 2 + 5$, so $x = 2$.

Example 2. If $3x = 7x$, then $x = 0$ or $3 = 7$. Of course $3 \neq 7$, so we must have $x = 0$.

We are now able to establish two other **cancellation properties**.

Theorem 4.7.7. *Assume $a, b, c \in W$. If $a + c < b + c$, then $a < b$.*

Theorem 4.7.8. *Assume $a, b, h \in W$. If $ah < bh$, then $a < b$.*

Proof of Theorem 4.7.7. If $a + c < b + c$, then there exists a positive whole number p such that $a + c + p = b + c$. But by the commutative and associative properties $a + c + p = a + p + c$, so we have $a + p + c = b + c$, and we can use Theorem 4.7.5 to cancel the c, giving $a + p = b$. Therefore, by definition, $a < b$.

Proof of Theorem 4.7.8. Here again we will find the contrapositive easier to prove than the given implication. We will prove that if $a \not< b$, then $ah \not< bh$. Now if $a \not< b$, then by the trichotomy property, $a = b$ or $b < a$. If $a = b$, then $ah = bh$ so $ah \not< bh$. If $b < a$, then unless $h = 0$ we know from Theorem 4.7.4 that $bh < ah$, and so by trichotomy $ah \not< bh$. Of course, if $h = 0$, then $ah = bh$, and we still have $ah \not< bh$. Hence we have shown that if $a \not< b$ then $ah \not< bh$. Therefore, if $ah < bh$, then $a < b$.

Exercises

1. Use the restatement of Theorem 4.7.6. to prove that, for whole numbers a and h, if $ah = 0$ and $h \neq 0$, then $a = 0$.
2. Prove that if p is a positive whole number then
   (a) $p < 2p$; (b) $5p < 7p$.
3. Prove that when b is a whole number and a is a positive whole number, $ab < a(b + 1)$.
4. Prove that for positive whole numbers a and b, if $a < b$ then
   (a) $a^2 < ab$; (b) $a^2 < b^2$.

4.8 Subtraction and division

Another useful symbol is "$\leq$". It means "less than or equal to," that is, $a \leq b$ means $a < b$ or $a = b$. Of course, by the trichotomy property, we cannot have both $a < b$ and $a = b$. Similarly, $r \geq s$ means $r > s$ or $r = s$. For example, $3 \leq 7$, $2 \leq 2$, $8 \geq 5$, and $7 \geq 7$.

If a and b are whole numbers and $a \leq b$, then there is a whole number x such that $a + x = b$. In fact, if $a < b$ there is a positive whole number x with the given property, and if $a = b$, then $a + 0 = b$. If $a > b$, however, there is no whole number x such that $a + x = b$, because $a + 0 = a \neq b$, and if p is positive and $a + p = b$, then $a < b$, so $a \not> b$.

Example 1. $4 \leq 7$ and $4 + 3 = 7$.

Example 2. $5 \leq 5$ and $5 + 0 = 5$.

Example 3. $4 > 3$ and there is no whole number x such that $4 + x = 3$.

Definition 4.8.1. (*Definition of subtraction of whole numbers.*) *If a and b are whole numbers, then $a - b$ (read a minus b) is the number x such that $b + x = a$.*

Another way to state this definition is to say that $a - b$ *is the number which when added to b produces a.* For example, $5 - 2 = 3$ because $2 + 3 = 5$, and $4 - 4 = 0$ because $4 + 0 = 4$.

As we have already shown, such a number x exists if and only if $a \geq b$, so $a - b$ does not have a meaning for numbers a and b in W when $a < b$. For example, $3 - 5$ has no meaning in W: W is not closed under subtraction.

In the definition of $a - b$ we have said that $a - b$ is *the* number x such that $b + x = a$. Could it be that there are two (or more) different numbers with this property? No, because if $b + x = a$ and $b + y = a$, then $b + x = b + y$, and hence $x + b = y + b$ and we can use Theorem 4.7.5 to cancel the b's to get $x = y$. Hence if there is any number x for which $b + x = a$, then there is a *unique* x with that property. This justifies saying *the* number x such that $b + x = a$.

Another thing to notice about the definition of $a - b$ is that since $a - b$ is the number x such that $b + x = a$, we have $b + (a - b) = a$. We will need this fact, in the form $c + (b - c) = b$, in the next theorem.

Also since $(a - b) + b = a$ by the commutative property of addition, a is b units to the right of $a - b$ on the number line.

Thus we can mechanically obtain $a - b$ on the number line by going b units to the *left* of point a to arrive at $a - b$.

For example, we can obtain $5 - 2$ by going 2 units to the left of 5 to arrive at the result 3.

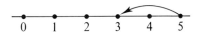

Theorem 4.8.1. $a(b - c) = ab - ac$ (*Multiplication is distributive over subtraction*).

Proof. By definition of $ab - ac$, we must prove that $a(b - c)$ is the number x such that $ac + x = ab$. That is, we must prove that

$$ac + a(b - c) = ab.$$

To do this, we use the distributive property to get

$$ac + a(b - c) = a[c + (b - c)].$$

Now, as we have mentioned above $c + (b - c) = b$, so

$$ac + a(b - c) = a[c + (b - c)] = ab.$$

Therefore, $a(b - c) = ab - ac$.

We now turn to division.

Definition 4.8.2. (*Definition of division of whole numbers.*) *If a and b are whole numbers, then $a \div b$ (read a divided by b) is the unique number x such that $bx = a$.*

Another way to state this definition is to say that *$a \div b$ is the number which when multiplied by b produces a.* For example:

    (1) $6 \div 2 = 3$ because 3 is the number x such that $2x = 6$.
    (2) $10 \div 5 = 2$ because $5 \cdot 2 = 10$.
    (3) $7 \div 1 = 7$ because $1 \cdot 7 = 7$.

In many cases, division has no meaning in W. For example, $3 \div 2$ has no meaning in W, because there is no whole number x such that $2x = 3$. Also, $5 \div 0$ has no meaning in W, because there is no number x such that $0 \cdot x = 5$. Of course, $0 \cdot x = 0$ no matter what x is, so $a \div 0$ can never have a meaning if $a \neq 0$, because then there is no number x such that $0 \cdot x = a$.

What about $0 \div 0$? It would have to be *the* number x such that $0 \cdot x = 0$. Here there is no *unique* number x with this property; in fact, $0 \cdot x = 0$ for every $x \in W$. Hence, $0 \div 0$ has no meaning, and consequently $a \div b$ never has a meaning when $b = 0$, no matter what number a is.

Notice that if $b \neq 0$ and there is in fact any number x in W for which $bx = a$, then there must be a *unique* number in W with this property, because if y is also a number such that $by = a$, then $bx = by$ and hence by Theorem 4.7.6 we can cancel the b to get $x = y$.

If $a \div b = c$, we call a the **dividend**, b the **divisor**, and c the **quotient**.

Note that subtraction and division are not commutative or associative. For example,

(1) $8 - 5 \neq 5 - 8$, so subtraction is not commutative.

(2) $8 \div 2 \neq 2 \div 8$, so division is not commutative.

(3) $(8 - 5) - 2 = 3 - 2 = 1$ while $8 - (5 - 2) = 8 - 3 = 5$, so subtraction is not associative.

(4) $(12 \div 6) \div 2 = 2 \div 2 = 1$ while $12 \div (6 \div 2) = 12 \div 3 = 4$, so division is not associative.

Another term that is especially important is **divides**.

Definition 4.8.3. (*Definition of divides, multiple, and divisor*) *If a and b are whole numbers and there exists a whole number x such that $bx = a$, we say that b **divides** a or that a is a **multiple** of b or that b is a **divisor** of a.*

We write $b|a$ for b divides a.

Examples:

(1) $2|6$ because there is a number $x = 3$ such that $2x = 6$, that is, $2 \cdot 3 = 6$. Also 2 and 1 are divisors of 6, and 6 is a multiple of 2 and a multiple of 1.

(2) $4|8$ because $4 \cdot 2 = 8$. Also 8 is a multiple of 4, 8 is a multiple of 2, and 2 and 4 are both divisors of 8.

(3) $3|6$ because $3 \cdot 2 = 6$.

(4) 2 does not divide 3, that is, $2 \nmid 3$, because there is no $x \in W$ such that $2x = 3$.

It is to be emphasized that $b|a$ and $a \div b$ mean different things. The first is a relationship between a and b while the second, $a \div b$, is a number. In particular, $2|6$ because $6 \div 2$ exists in W. Of course, $6 \div 2$ is the number 3.

Because $a - b$ and $a \div b$ are not always meaningful in W, we will later develop larger systems of numbers containing W in which $a - b$ and $a \div b$ always exist (with the exception that $a \div b$ will never exist when $b = 0$).

We learned in Section 4.5 that a product cd of whole numbers c and d can be found on the number line by going c steps, of d units each, to the right of the origin. We can apply this to division, when b divides a, by noting that by the definition of division, $b(a \div b) = a$; then, by the commutative property of multiplication, $(a \div b)b$, which is a, is obtained by going $a \div b$ steps, of b units each, to the right of the origin. Consequently, we can find $a \div b$ on the number line by determining how many steps, of b units each, to the right are required to go from the origin to a.

For example, to find $6 \div 2$ using the number line, we determine how many steps, of two units each, to the right are needed to go from the origin to 6. We see that 3 such steps are needed and thus obtain the result 3.

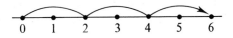

Exercises

1. Decide whether each of the following exists in W and determine its value if it does exist.

   (a) $10 \div 4$;   (b) $26 \div 13$;   (c) $91 \div 7$;   (d) $0 \div 2$;
   (e) $2 \div 0$;   (f) $15 - 6$;   (g) $0 - 2$;   (h) $2 - 0$;
   (i) $0 - 0$;   (j) $0 \div 0$;   (k) $3 \div 1$;   (l) $1 \div 3$;
   (m) $10 - (3 - 2)$;   (n) $(10 - 3) - 2$;   (o) $(36 \div 6) \div 2$;   (p) $36 \div (6 \div 2)$.

2. Draw appropriate number lines and use them to determine each of the following:

   (a) $6 - 2$;   (b) $7 - 3$;   (c) $3 - 3$;   (d) $6 \div 3$;
   (e) $8 \div 2$;   (f) $9 \div 3$;   (g) $4 \div 4$;   (h) $10 \div 2$.

3. Is it true that $7|0$? that $0|0$? that $4|10$?

4. Prove that if $b|a$, then $b|ac$.

5. Prove that if $d|a$ and $d|b$, then $d|(a + b)$.

6. Show that if $d|ab$ then d does not have to divide a or b, by finding specific positive whole numbers a, b, d such that $d|ab$ but $d \nmid a$ and $d \nmid b$.

7. Explain why it *follows directly from the definition* of subtraction of whole numbers that

   (a) $b + (a - b) = a$;   (b) $(a + b) - a = b$.

8. Explain why it *follows directly from the definition* of division of whole numbers that

   (a) $b(a \div b) = a$;   (b) $(ab) \div a = b$.

4.9 The division algorithm

We begin this section with a detailed numerical example which contains statements that would be unnecessary if we were interested only in this example and not in the general situation.

Let us take the numbers 7 and 3 and notice that $3q$ increases when q increases. When $q = 2$, $3q = 6 < 7$, and $3(q + 1) = 3 \cdot 3 = 9 > 7$. Thus $q = 2$ is the unique largest whole number q for which $3q \leq 7$. Now, since $3 \cdot 2 \leq 7$, $7 = 3 \cdot 2 + r$, where r is unique and $0 \leq r$. Also, $r < 3$, for if $r \geq 3$ then $3 \cdot 2$ is not the largest multiple of 3 that is less than or equal to 7. In this case $r = 1$ and $7 = 3 \cdot 2 + 1$ where $0 \leq 1 < 3$.

More generally, using a instead of 7 and b instead of 3, there is a unique largest multiple bq of b that is less than or equal to a. Now, since $bq \leq a$, $a = bq + r$ where r is unique and $0 \leq r$. Also, $r < b$, for if $r \geq b$ then bq is not the largest multiple of b that is less than or equal to a. Therefore, $a = bq + r$ where $0 \leq r < b$. This leads to the next theorem known as the **division algorithm**.

Theorem 4.9.1. (*The division algorithm.*) *If a and b are whole numbers and $b \neq 0$, then there exist unique whole numbers q and r such that $a = bq + r$ where $0 \leq r < b$.*

Note that we do not require that $b < a$. In the relationship $a = bq + r$, a is called the **dividend**, b the **divisor**, q the **quotient**, and r is called the **remainder**.

Examples:

(1) $a = 17$, $b = 4$, $17 = 4 \cdot 4 + 1$, $0 \leq 1 < 4$.
(2) $a = 19$, $b = 4$, $19 = 4 \cdot 4 + 3$, $0 \leq 3 < 4$.
(3) $a = 22$, $b = 5$, $22 = 5 \cdot 4 + 2$, $0 \leq 2 < 5$.
(4) $a = 5$, $b = 22$, $5 = 22 \cdot 0 + 5$, $0 \leq 5 < 22$.
(5) $a = 35$, $b = 5$ $35 = 5 \cdot 7 + 0$, $0 \leq 0 < 5$.

Exercises

For each of the following values of a and b, find the unique whole numbers q and r such that $a = bq + r$ where $0 \leq r < b$.

1. $a = 213$, $b = 10$;   2. $a = 2$, $b = 9$;   3. $a = 0$, $b = 7$;
4. $a = 29$, $b = 6$;   5. $a = 17$, $b = 108$;   6. $a = 624$; $b = 7$.
7. Find the whole numbers a_0, a_1, a_2 such that $57 = a_2(5)^2 + a_1(5) + a_0$, where each of the a's is one of the numbers 0, 1, 2, 3, 4. (*Hint:* Find q and r such that $57 = 5q + r$ and then find t and u such that, for the q you have found, $q = 5t + u$. Then combine the two relations.)

chapter 5

Numerals

5.1 Early and present numeral systems

There is a difference between a number and the symbol or name used for a number, just as there is a difference between a person and the person's name. The symbol used for a number is called a **numeral**.

We do not know what the first numerals used by man looked like, but they were probably just lines drawn in the dirt or scratches on stone, such as ||||| for six. They may have been used for tallying animals or people, and someone would put the set he was tallying in 1–1 correspondence with the marks he was making. It is likely that the evolution proceeded when people found that this tallying could be read more easily if they made the marks in bunches such as ||||| ||||| || for twelve or as we often do today when we are tallying and write ⊬⊬⊬ ⊬⊬⊬ || for twelve. Because people in those days probably also "counted on their fingers," each bunch most likely contained five or ten marks. The next evolutionary step was to use some special separate symbol for each of the bunches.

The earliest numeral system for which we have evidence is of this type. It appears in Egyptian hieroglyphics which date back more than 5,000 years to about 3,400 B.C. This system used the symbol | for one, the symbol ∩ for ten, and ⑨ for one hundred (ten tens), as well as other symbols for 1,000; 10,000; etc. A number like 245 was cut in stone similarly to

The number 52 was expressed something like

Another known early form of numerals is the Babylonian cuneiforms which date to about 2,000 B.C. The cuneiforms were impressions made in wet clay and hardened in the sun. They used only two symbols | for one and ⟨ for ten. The symbol | was also used for 60, 60^2, and the higher powers of 60. Similarly ⟨ was used also for 10(60), $10(60)^2$, and 10 times higher powers of 60. Because of these ambiguities, it was sometimes difficult to know which number was being expressed. Hopefully, one somehow determined this from the context.

The cuneiform numerals for numbers up to 59 were similar to Egyptian numerals in that they used repetition and addition properties. For example, they represented 32 as ⟨⟨⟨ ||. Beyond 60 they used a combination of this and positional notation where the order of the symbols was important. We use a positional notation in our own familiar (Hindu-Arabic) numeral system. However, the Babylonian system was based on 60 and ours is based on 10. For example, 232 means $2(10)^2 + 3(10) + 2$ in our system. The position of each of the 2's has a great deal to do with its contribution to the number represented. The "place value" of the first 2 is 200 while the second one has a "place value" of 2. The Babylonian numeral ⟨|⟨⟨||, which is eleven followed by twenty-two, represents $11(60) + 22 = 682$.

The Roman numeral system which started sometime before 300 B.C. is familiar to us. It uses symbols I for 1, V for 5, X for 10, L for 50, C for 100, D for 500, and M for 1,000. The early forms used the symbols I, II, III, IIII for the numbers from 1 to 4. They later used a subtractive technique to express 4 as IV, 9 as IX, 40 as XL, 90 as XC, 400 as CD, and 900 as CM. In each of these cases, the order of the symbols is important. Each of them has a numeral representing a number preceded by a numeral representing a smaller number. In the system, this means subtraction. It is like reading the clock and saying the time is 5 minutes before 6 to see, for example, that XL is 10 before 50.

The Roman numeral system uses repetition and addition of the symbols (and sometimes subtraction). As other examples, MCMXXXIX represents 1,939 while CCCXXIV is the symbol for 324.

As mentioned previously, the numeral system we commonly use is called the Hindu-Arabic system. The reason for the name is that it was originated by the Hindus and introduced in Europe by the Arabs. Although the system was developed sometime before the birth of Christ, it was not until about 800 A.D. that the zero appeared. The techniques we use to add, subtract, multiply, and divide were not fully developed until about 1000 A.D. and did not come into extensive use in Europe until about 1500 A.D.

From now on we will use the convention that when we are indicating a numeral we will use quotation marks. For example, when we write 25 we will mean the number, but when we write "25" we will mean the numeral. Thus 2 is a number but "2" is not a number, it is numeral, Similarly, George may be 6 feet tall, but "George" is a six-letter word.

Our Hindu-Arabic numeral system is ingenious. It is so systematic and concise that it is much easier to read, write, and compute with it than with any of the other systems described above.

We need only ten symbols in our system, namely, "0, 1, 2, 3, 4, 5, 6, 7, 8, 9," and all the whole numbers can be expressed conveniently using combinations of these ten symbols. The number ten is represented by "10."

When we write "23," it does not represent 2×3 or $2 + 3$ but $2 \cdot (10) + 3$. Also, "6434" is the numeral for $6(10)^3 + 4(10)^2 + 3(10) + 4$. The position of each symbol is important because each one has a certain "place value." In "6434" the "4" on the right represents 4, but the other "4" represents 400,

Every one of our numerals has the form "$a_h \ldots a_2 a_1 a_0$" where the a's are called **digits** and each one is one of the numerals "0, 1, 2, 3, 4, 5, 6, 7, 8, 9." The symbol "$a_h \ldots a_2 a_1 a_0$" stands for the number

$$a_h(10)^h + \ldots + a_2(10)^2 + a_1(10) + a_0,$$

and "a_0" is called the units digit, "a_1" the tens digit, "a_2" the hundreds digit, etc.

Every number in W can be expressed in this way as a sum of terms, each of which is 10 to some exponent times one of the numbers $0, 1, 2, \ldots, 9$. Ten is the **base** of our numeral system, which is often called the **decimal** numeral system. Actually, any number greater than 1 could be used for the base of a similar numeral system.

Exercises

Express each of the following numbers in our familiar Hindu-Arabic system:

1. ||| ∩∩ 9 ;
2. |||| ∩∩∩ 999 ;
   ||| ∩∩∩
3. <<<||.
    << ||'
4. <<< |< ||| ;
5. MMCMXLIII.

Express the following numbers in each of these systems; (a) Egyptian hieroglyphics, (b) Babylonian, and (c) Roman:

6. 17;   7. 26;   8. 39;   9. 99.

5.2   Hindu-Arabic numeral systems to other bases

In order to understand our system with base 10 better and to see what systems with other bases are like, we will spend some time considering numeral systems with base 5, base 3, and base 12.

In the base 5 system, every whole number is expressed in the form

$$b_k(5)^k + \cdots + b_2(5)^2 + b_1(5) + b_0$$

where each of the b's is one of the numbers 0, 1, 2, 3, 4, which is written more simply as "$(b_k \cdots b_2 b_1 b_0)_5$." The parentheses and the subscript 5 are used so that we do not confuse it with a numeral of our familiar base 10 system. When no subscript appears, the base is 10. For example, ten in the base 5 system is $(20)_5 = 2(5) + 0$, and $(341)_5 = 3(5)^2 + 4(5) + 1 = 3(25) + 20 + 1 = 96$. Also $(10)_5 = 1(5) + 0 = 5$ which, of course, is five and not ten.

In order to see how to express a number in the base 5 system, let us look at $96 = (341)_5 = 3(5)^2 + 4(5) + 1$. How would we get the digits "3," "4," and "1" if we were changing 96 to the base 5 system? Notice that $96 = [3(5) + 4]5 + 1$, where we have used the distributive property to get $3(5)^2 + 4(5) = [3(5) + 4]5$. Hence, the 1 is the unique remainder r obtained when we express $96 = 5q + r$, where $0 \le r < 5$. Here the $q = 3(5) + 4$ so the 4 is the unique number r_1 obtained when we write $q = 5q_1 + r_1$, where $q_1 = 3$ and $0 \le r_1 < 5$. Now q_1 is the 3; if q_1 were not in the interval $0 \le q_1 < 5$ though, we would express $q_1 = 5q_2 + r_2$, where $0 \le r_2 < 5$, and r_2 would be the next digit, etc.

In the general case, to express any whole number n in the base 5 system, we write

$$
\begin{array}{llll}
n = 5q & + r_0 & \text{where} & 0 \le r_0 < 5, \\
q = 5q_1 & + r_1 & \text{where} & 0 \le r_1 < 5 \\
q_1 = 5q_2 & + r_2 & \text{where} & 0 \le r_2 < 5, \\
q_2 = 5q_3 & + r_3 & \text{where} & 0 \le r_3 < 5,
\end{array}
$$

etc., until we get to q_k such that $0 \le q_k < 5$. Then $q_k = 5(0) + r_k$ where $0 \le r_k < 5$, and $n = (r_k \ldots r_2 r_1 r_0)_5$.

Let us use 184 as an example:

$$
\begin{array}{ll}
184 = 5(36) & + 4, \\
36 = 5(7) & + 1, \\
7 = 5(1) & + 2, \\
1 = 5(0) & + 1.
\end{array}
$$

So $184 = (1214)_5$, and we could check our work by calculating.

$$(1214)_5 = 1(5)^3 + 2(5)^2 + 1(5) + 4 = 184.$$

In the base 3 system every number of W is expressed in the form

$$C_s(3)^s + \ldots + c_2(3)^2 + c_1(3) + C_0$$

where each of the c's is one of the numbers 0, 1, 2 and we write it more simply as "$(c_s \ldots c_2 c_1 c_0)_3$." For example, 10 in the base 3 system is $1(3)^2 + 0(3) + 1 = (101)_3$. The number $(10)_3 = 1(3) + 0 = 3$ and should not be read as "ten" but as "one-oh base 3", and could also be read as "three." Also.

$$\begin{aligned}
(21012)_3 &= 2(3)^4 + 1(3)^3 + 0(3)^2 + 1(3) + 2 \\
&= 2(81) + 1(27) + 1(3) + 2 = 194.
\end{aligned}$$

Similarly to the base 5 system, to express a whole number n in the base 3 system we write

$$\begin{aligned}
n &= 3q + r_0 \qquad \text{where} \qquad 0 \le r_0 < 3 \\
q &= 3q_1 + r_1 \qquad \text{where} \qquad 0 \le r_1 < 3 \\
q_1 &= 3q_2 + r_2 \qquad \text{where} \qquad 0 \le r_2 < 3 \\
q_2 &= 3q_3 + r_3 \qquad \text{where} \qquad 0 \le r_3 < 3
\end{aligned}$$

etc., until we get to a q_s such that $0 \le q_s < 3$. Then $q_s = 3(0) + r_s$ where $0 \le r_s < 3$, and $n = (r_s \ldots r_2 r_1 r_0)_3$.

As an example, consider the change of 107 to the base 3 system.

$$\begin{aligned}
107 &= 3(35) + 2 \\
35 &= 3(11) + 2 \\
11 &= 3(3) + 2 \\
3 &= 3(1) + 0 \\
1 &= 3(0) + 1
\end{aligned}$$

So $107 = (10222)_3$ and we could check the work by calculating

$$(10222)_3 = 1(3)^4 + 0(3)^3 + 2(3)^2 + 2(3) + 2 = 107.$$

In the base 12 system, twelve symbols are needed, and hence we will have to use two symbols in addition to the ten familiar ones. We will use "0, 1, 2, 3, 4, 5, 6, 7, 8, 9, T, E," where the new symbols "T" and "E" are used for ten and eleven respectively. Every number of W is expressed in the form

$$d_h(12)^h + \ldots + d_2(12)^2 + d_1(12) + d_0$$

where each d is one of the numbers 0, 1, 2, 3, 4, 5, 6, 7, 8, 9, T, E, It is written more simply as "$(d_h \ldots d_2 d_1 d_0)_{12}$." For example,

$$(20T3E)_{12} = 2(12)^4 + 0(12)^3 + 10(12)^2 + 3(12) + 11$$
$$= 2(20736) + 10(144) + 36 + 11 = 42{,}959.$$

Also, we change 2147 to the base 12 system as follows:

$$\begin{array}{llll}
2147 = 12(178) + 11 & \text{where} & 0 \leq 11 < 12 \\
178 = 12(14) + 10 & \text{where} & 0 \leq 10 < 12 \\
14 = 12(1) + 2 & \text{where} & 0 \leq 2 < 12 \\
1 = 12(0) + 1 & \text{where} & 0 \leq 1 < 12
\end{array}$$

Then $2147 = (12TE)_{12}$.

It is important to remember that all of the familiar word names, "one," "two," "three," "four," etc., of the numbers remain the same no matter what numeral system is used. For example, 15, XV, $(30)_5$, $(120)_3$, and $(13)_{12}$ are all equal and have the same word name, "fifteen." The word names are especially useful when calculating with one digit numbers when the base is not 10. For example, to find $(3)_5 + (4)_5$ the simplest way is to think "three plus four is seven, and seven in the base 5 system is $(12)_5$." Similarly to find $(4)_{12} \times (5)_{12}$ the simplest way is to think "four times five is twenty, and twenty in the base 12 system is $(18)_{12}$."

Exercises

1. Express each of the numbers from 0 to 32 in the base 5 system, the base 3 system, and the base 12 system.

2. Express 239 and 786 in the base 5 system, the base 3 system, and the base 12 system.

3. Change $(4132)_5$ to the base 10 system.

4. Change $(12012)_3$ to the base 10 system.

5. Express 79 in the base 4 system.

6. Change $(5T8E)_{12}$ to the base 10 system.

7. Use the word names of the numbers to help find each of the following (each answer should be expressed in the base system given):

    a) $(3)_5 + (3)_5$;   b) $(3)_5 \times (4)_5$;   c) $(2)_5 \times (3)_5$;

    d) $(4)_5 + (4)_5$;   e) $(2)_5 + (3)_5$;   f) $(4)_5 \times (4)_5$;

    g) $(2)_3 + (2)_3$;   h) $(2)_3 \times (2)_3$;   i) $(7)_{12} + (5)_{12}$;

    j) $(9)_{12} + (7)_{12}$;   k) $(7)_{12} \times (2)_{12}$;   l) $(T)_{12} + (E)_{12}$.

5.3   Addition in our numeral system

We will now examine the familiar process used when we add whole numbers expressed in our Hindu-Arabic numeral system and see why this process is correct. To achieve a better understanding of it, we will first consider the addition process in the base 5 system in detail.

Consider the specific example $(213)_5 + (121)_5$.

$$(213)_5 + (121)_5 = 2(5)^2 + 1(5) + 3 + 1(5)^2 + 2(5) + 1$$
$$= 2(5)^2 + 1(5)^2 + 1(5) + 2(5) + (3 + 1)$$

by the commutative and associative properties.

Now by the distributive property, we see that we can write this as

$$(2 + 1)(5)^2 + (1 + 2)(5) + (3 + 1) = 3(5)^2 + 3(5) + 4 = (334)_5.$$

An important part of this is that

$$(213)_5 + (121)_5 = (2+1)(5)^2 + (1+2)5 + (3+1).$$
$$= 3(5)^2 + 3(5) + 4 = (334)_5.$$

Each digit of the first numeral is added to the corresponding digit of the second numeral. It follows, in a similar fashion, that to add any two numbers in the base 5 system or base 10 system or a system with any base, one adds corresponding digits. Because of this we can conveniently perform the addition in columns as follows, where we have also written the calculation in terms of the meaning of the numerals.

$$
\begin{array}{ll}
2(5)^2 + 1(5) + 3 & (213)_5 \\
1(5)^2 + 2(5) + 1 & (121)_5 \\
\hline
3(5)^2 + 3(5) + 4 & (334)_5
\end{array}
$$

But it is not always this simple. For example, take $(434)_5 + (324)_5$. We will write out each step of this calculation in terms of the meaning of the numerals and also using just the numerals. We begin with the column on the right and work toward the left, one step at a time.

$$
\begin{array}{ll}
4(5)^2 + 3(5) + 4 & (434)_5 \\
3(5)^2 + 2(5) + 4 & (324)_5 \\
\hline
\end{array}
$$

Now $4 + 4 = 1(5) + 3$, so we get ("carrying" the 1, that is, adding the $1(5)$ with the other multiples of 5)

$$
\begin{array}{cc}
1(5) & 1 \\
4(5)^2 + 3(5) + 4 & (434)_5 \\
3(5)^2 + 2(5) + 4 & (324)_5 \\
\hline
+ 3 & (\quad 3)_5
\end{array}
$$

Now $1(5) + 3(5) + 2(5) = (1 + 3 + 2)(5) = (1(5) + 1)5 = 1(5)^2 + 1(5)$ so we again "carry" the 1, that is, we add the $1(5)^2$ with the other multiples of 5^2, to obtain

$$
\begin{array}{cc}
1(5)^2 \quad 1(5) & 11 \\
4(5)^2 + 3(5) + 4 & (434)_5 \\
3(5)^2 + 2(5) + 4 & (324)_5 \\
\hline
1(5) + 3 & (\ 13)_5
\end{array}
$$

Then $(1 + 4 + 3)(5)^2 = (1(5) + 3)(5)^2 = 1(5)^3 + 3(5)^2$ so the next step gives us

$$
\begin{array}{cc}
1(5)^3 \quad 1(5)^2 \quad 1(5) & 111 \\
4(5)^2 + 3(5) + 4 & (434)_5 \\
3(5)^2 + (52) + 4 & (324)_5 \\
\hline
3(5)^2 + 1(5) + 3 & (313)_5
\end{array}
$$

And finally we finish with

$$
\begin{array}{cc}
1(5)^3 \quad 1(5)^2 \quad 1(5) & 111 \\
4(5)^2 + 3(5) + 4 & (434)_5 \\
3(5)^2 + 2(5) + 4 & (324)_5 \\
\hline
1(5)^3 + 3(5)^2 + 1(5) + 3 & (1313)_5.
\end{array}
$$

So $(434)_5 + (324)_5 = (1313)_5$.

An example in the base 3 system carried out in terms of the meaning of the numerals and also using just the numerals is as follows:

$$
\begin{array}{cc}
1(3)^4 \qquad 1(3)^2 \quad 1(3) & 1 \ 11 \\
1(3)^3 + 1(3)^2 + 1(3) + 2 & (1112)_3 \\
2(3)^3 + 0(3)^2 + 2(3) + 2 & (2022)_3 \\
\hline
1(3)^4 + 0(3)^3 + 2(3)^2 + 1(3) + 1 & (10211)_3
\end{array}
$$

Therefore $(1112)_3 + (2022)_3 = (10211)_3$.

A similar example in the base 10 system with all the steps done is:

$$
\begin{array}{ll}
1(10)^3 \qquad\qquad 1(10) & 1\ 1 \\
6(10)^3 + 5(10)^2 + 4(10) + 7 & 6547 \\
\qquad\qquad 7(10)^2 + 0(10) + 8 & \underline{708} \\
\hline
7(10)^3 + 2(10)^2 + 5(10) + 5 & 7255
\end{array}
$$

In order to use the familiar process described above for the base 10 system, one should memorize (as all of us have already done) the "addition table" for sums of the form $a + b$ where a and b are any of the numbers 0, 1, 2, 3, 4, 5, 6, 7, 8, 9. As we have pointed out in the section covering the definition of addition, we can prove, by using the appropriate sets and the definition of addition, that these sums (such as $2 + 2 = 4$) that we memorized years ago are correct.

If one were doing much addition using the base 3 or base 5 systems he would memorize the addition table of one digit sums for that base. The one-digit addition tables for the base 5 and base 3 systems are as follows:

Base 5 addition table

+	1	2	3	4
1	2	3	4	10
2	3	4	10	11
3	4	10	11	12
4	10	11	12	13

Base 3 addition table

+	1	2
1	2	10
2	10	11

For example, in the base 5 system, $2 + 3$ is the 10 which appears in the same row as the 2 on the left side of the table and in the same column as the 3 at the top of the table.

Exercises

Find the following sums, showing the calculation first in terms of the meaning of the numerals and then using just the numerals.

1. $(14303)_5 + (2434)_5$; 2. $(3241)_5 + (4344)_5$;

3. $(21120)_3 + (10221)_3$; 4. $(12022)_3 + (2102)_3$;

5. $78258 + 3794$; 6. $3679 + 8752$.

7. Find the one-digit addition table for the base 4 system, and the base 12 system.

5.4   Subtraction in our numeral system

Let us now see why the familiar subtraction process in our numeral system is correct. We will consider this in the base 10 system, in the base 5 system, and in the base 3 system.

First take a simple example, $95 - 61$. We get

$$95 - 61 = [9(10) + 5] - [6(10) + 1].$$

We say that, by the definition of subtraction, this equals.

$$[9(10) - 6(10)] + (5 - 1),$$

because when we add $6(10) + 1$ to this last expression we get $9(10) + 5$, as we shall now show. By the commutative and associative properties, we get $6(10) + 1 + [9(10) - 6(10)] + (5 - 1) = 6(10) + [9(10) - 6(10)] + 1 + (5 - 1)$. But by the definition of subtraction, $6(10) + [9(10) - 6(10)] = 9(10)$ and $1 + (5 - 1) = 5$, so $6(10) + 1 + [9(10) - 6(10)] + (5 - 1) = 9(10) + 5$ which is what we wanted to show. Therefore,

$$95 - 61 = [9(10) + 5] - [6(10) + 1] = [9(10) - 6(10)] + (5 - 1).$$

But by the distributive property (Theorem 4.3.2), we have $9(10) - 6(10) = (9 - 6)(10)$, so

$$95 - 61 = [9(10) - 6(10)] + (5 - 1) = (9 - 6)(10) + (5 - 1) = 3(10) + 4 = 34.$$

As in the case of addition, an important part of this is that

$$95 - 61 = (9 - 6)(10) + (5 - 1).$$

In the same way

$$874 - 352 = (8 - 3)(10)^2 + (7 - 5)(10) + (4 - 2).$$

In similar fashion, it follows that to subtract one number from another number, one subtracts corresponding digits. Because of this, we can conveniently do the subtraction in columns as follows, where we write the calculation in terms of the meaning of the numerals and also just using the numerals.

$$\begin{array}{ll}
8(10)^2 + 7(10) + 4 & \quad 874 \\
3(10)^2 + 5(10) + 2 & \quad 352 \\
\hline
5(10)^2 + 2(10) + 2 & \quad 522
\end{array}$$

But it can be more complicated than this; for example, take $953 - 678$. As with addition, we begin on the right and work toward the left. We proceed 1 step at a time as follows, writing each step in two ways.

$$9(10)^2 + 5(10) + 3 \qquad 953$$
$$6(10)^2 + 7(10) + 8 \qquad 678$$

Now we cannot have $3 - 8$ in W, so we write the $5(10) + 3$ as $4(10) + (1(10) + 3) = 4(10) + 13$ and subtract $13 - 8$ to get 5. We have thus "borrowed" to change $5(10)$ to $4(10)$ and 3 to 13, giving

$$9(10)^2 + 4(10) + 13 \qquad 94^|3$$
$$6(10)^2 + 7(10) + \ 8 \qquad 67\ 8$$
$$\overline{\hphantom{9(10)^2 + 4(10) +} 5} \qquad \overline{\hphantom{67} 5}$$

For the usual shortened way of writing the calculation illustrated on the right, we have changed 953 to $94^|3$. The small 1 written between the tops of the 4 and 3 indicates that we have changed the 3 to 13. Again we cannot have $4 - 7$ so we borrow again, using the fact that $9(10)^2 + 4(10) = 8(10)^2 + [1(10)^2 + 4(10)] = 8(10)^2 + [1(10) + 4]10 = 8(10)^2 + 14(10)$, which gives

$$8(10)^2 + 14(10) + 13 \qquad 8^|4^|3$$
$$6(10)^2 + \ 7(10) + \ 8 \qquad 6\ 7\ 8$$
$$\overline{\hphantom{8(10)^2 +} 7(10) + \ 5} \qquad \overline{\hphantom{6} 7\ 5}$$

Thus the illustrations show that we have changed $9(10)^2 + 4(10)$ to $8(10)^2 + 14(10)$ and in the shortened form we have changed the 943 to $8^|4^|3$. As before, the $^|4$ shows the change from 4 to 14 obtained by borrowing 1 from the 9. We finally get

$$8(10)^2 + 14(10) + 13 \qquad 8^|4^|3$$
$$6(10)^2 + \ 7(10) + \ 8 \qquad 6\ 7\ 8$$
$$\overline{2(10)^2 + \ 7(10) + \ 5} \qquad \overline{2\ 7\ 5}$$

Consequently $953 - 678 = 275$.

As another example, take $7482 - 3816$. This is done in the following steps:

(1) 7482	(2) $747^|2$	(3) $747^|2$	(4) $6^|47^|2$	(5) $6^|47^|2$
3816	381 6	381 6	3 81 6	3 81 6
——	——	——	——	——
6	6	6 6	66 6	3 66 6

So $7482 - 3816 = 3666$.

Usually one performs these borrowing changes only mentally, so that the completed work looks like this:

$$7482$$
$$3816$$
$$\overline{3666}$$

In order to use more easily the familiar process described above to perform the operation of subtraction, in elementary school we memorized the results of subtractions of the form $a - b$ where a is a number from 1 to 19 (we need to go beyond 9 to allow for cases where borrowing is necessary) and b is a number from 1 to 9. We thus memorized a "subtraction table." The fact that these subtractions are actually correct can be proved from the definition of subtraction.

Now let us do an example in the base 5 system, one step at a time, first indicating what is happening according to the meaning of the numerals. Take $(321)_5 - (134)_5$.

$$
\begin{array}{ll}
3(5)^2 + 2(5) + 1 & (321)_5 \\
1(5)^2 + 3(5) + 4 & (134)_5 \\
\hline
\end{array}
$$

Then, because $(11)_5 - (4)_5 = (2)_5$, we get

$$
\begin{array}{ll}
3(5)^2 + 1(5) + 11 & (31^{|}1)_5 \\
1(5)^2 + 3(5) + 4 & (13\ 4)_5 \\
\hline
2 & (2)_5
\end{array}
$$

Now because $(11)_5 - (3)_5 = (3)_5$ we get

$$
\begin{array}{ll}
2(5)^2 + 11(5) + 11 & (2^{|}1^{|}1)_5 \\
1(5)^2 + 3(5) + 4 & (1\ 3\ 4)_5 \\
\hline
3(5) + 2 & (3\ 2)_5
\end{array}
$$

Finally we have

$$
\begin{array}{ll}
2(5)^2 + 11(5) + 11 & (2^{|}1^{|}1)_5 \\
1(5)^2 + 3(5) + 4 & (1\ 3\ 4)_5 \\
\hline
1(5)^2 + 3(5) + 2 & (1\ 3\ 2)_5
\end{array}
$$

So $(321)_5 - (134)_5 = (132)_5$.

With due care and understanding the borrowing can be done mentally so that the completed work can be simply

$$(321)_5$$
$$(134)_5$$
$$\overline{(132)_5}$$

Similarly in the base 3 system the completed work for $(21211)_3 - (10122)_3$ could be just

$$(21211)_3$$
$$(10122)_3$$
$$\overline{(11012)_3}$$

In order to subtract more easily in the base 3 system, one could use the base 3 " subtraction table ":

$-$	1	2
12	11	10
11	10	2
10	2	1
2	1	0
1	0	

In this table, all the numerals are in the base 3 system with $(\ )_3$ left off for convenience. Notice that there is a blank space for $1 - 2$ because it is not in W.

It is not recommended that this table be memorized because our aim is understanding, not speed in calculation.

A simple way to obtain all differences shown in the subtraction table is to use a number line labeled with the numerals of whatever base system is under consideration. In the base 3 system we have

$$(0)_3 \quad (1)_3 \quad (2)_3 \quad (10)_3 \quad (11)_3 \quad (12)_3$$

For example, to subtract $(11)_3 - (2)_3$ go 2 units to the left of $(11)_3$, thus arriving at $(2)_3$. Hence $(11)_3 - (2)_3 = (2)_3$.

Similarly in the base 5 system we have

$$(0)_5 \quad (1)_5 \quad (2)_5 \quad (3)_5 \quad (4)_5 \quad (10)_5 \quad (11)_5 \quad (12)_5 \quad (13)_5 \quad (14)_5$$

To subtract say $(12)_5 - (4)_5$, go 4 units to the left of $(12)_5$ to get $(3)_5$. Hence $(12)_5 - (4)_5 = (3)_5$.

Exercises

1. Draw a number line and label it with base 7 numerals from $(0)_7$ to $(16)_7$. Use this to find the following differences

(a) $(15)_7 - (6)_7$; (b) $(12)_7 - (5)_7$; (c) $(11)_7 - (6)_7$.

Find the following differences, showing the calculation first in terms of the meaning of the numerals and then using just the numerals.

2. $7423 - 5784$;

3. $5109 - 349$;

4. $(4021)_5 - (3344)_5$;

5. $(14211)_5 - (4342)_5$;

6. $(21201)_3 - (12122)_3$;

7. $(12001)_3 - (2112)_3$.

5.5   Multiplication in our numeral system

Again we will consider the base 10 system first. In order to multiply any two numbers in the base 10 system easily, we must know the "multiplication table" for products of the form $a \times b$ where a and b are any of the numbers 0, 1, 2, 3, 4, 5, 6, 7, 8, 9; that is, we must know the one-digit multiplication table.

It is also necessary to know how to multiply a number by 10 or 10^2 or 10^3 or 10^k for any $k > 3$. Now, for example, $724 \times 10 = [7(10)^2 + 2(10) + 4]10$, which by the distributive property equals

$$7(10)^3 + 2(10)^2 + 4(10) = 7(10)^3 + 2(10)^2 + 4(10) + 0 = 7240.$$

Also,

$$724 \times 10^2 = [7(10)^2 + 2(10) + 4](10)^2 = 7(10)^4 + 2(10)^3 + 4(10)^2$$
$$= 7(10)^4 + 2(10)^3 + 4(10)^2 + 0(10) + 0 = 72,400.$$

Similarly,

$$724 \times 10^3 = 724,000.$$

In general to multiply a whole number by 10^k, we just put k zeros on the end of the number; for example, $386 \times 10^5 = 38,600,000$.

We shall first consider products where one of the factors is a one-digit number. For example, $823 \times 4 = [8(10)^2 + 2(10) + 3]4$. By the distributive property, this equals.

$$(8 \cdot 4)(10)^2 + (2 \cdot 4)(10) + 3 \cdot 4.$$

Now $(8 \cdot 4)(10)^2$ ends with two zeros and hence does not affect the units or tens digits, and $(2 \cdot 4)(10)$ ends in 0 and so does not affect the units digit. This makes the addition of these three numbers very simple; we only need to be careful to "carry" when necessary. Again we work from right to left, first calculating $3 \cdot 4$, then adding $(2 \cdot 4)(10)$, then adding $(8 \cdot 4)(10)^2$. It is convenient to write the work as follows, where we indicate the steps one at a time in both the usual short form and the expanded form which shows the meaning of the base 10 notation. Now $3 \cdot 4 = 12 = 1(10) + 2$. We write $1(10)$ in the 10 column and 2 in the unit column.

$$
\begin{array}{rr}
1(10) & 1 \\
8(10)^2 + 2(10) + 3 & 823 \\
4 & 4 \\
\hline
2 & 2
\end{array}
$$

We have "carried" the $1(10)$ to be added in with the next step, where we first multiply $2 \cdot 4(10)$. We get $8(10) + 1(10) = 9(10)$.

$$
\begin{array}{rr}
1(10) & 1 \\
8(10)^2 + 2(10) + 3 & 823 \\
4 & 4 \\
\hline
9(10) + 2 & 92
\end{array}
$$

The next step is to add in the $8 \cdot 4(10)^2$, which is equal to $32 \times 10^2 = [3(10) + 2](10)^2 = 3(10)^3 + 2(10)^2$.

$$
\begin{array}{ccc}
3(10)^3 & 1(10) & 1 \\
& 8(10)^2 + 2(10) + 3 & 823 \\
& 4 & 4 \\
\hline
& 2(10)^2 + 9(10) + 2 & 292
\end{array}
$$

Again we have had to carry the $3(10)^3$. We then have the final step which gives us

$$
\begin{array}{ccc}
3(10)^3 & 1(10) & 3 \ 1 \\
& 8(10)^2 + 2(10) + 3 & 823 \\
& 4 & 4 \\
\hline
3(10)^3 + 2(10)^2 + 9(10) + 2 & 3292
\end{array}
$$

So $823 \times 4 = 3292$.

Similarly for any product of this type where one of the factors is a one digit number, we multiply the one-digit number by each of the digits of the other factor from right to left, "carrying" when necessary. Another example, 6×7438, with all the steps completed is as follows:

$$
\begin{array}{ccccc}
4(10)^4 & 2(10)^3 & 2(10)^2 & 4(10) & 4224 \\
& 7(10)^3 + 4(10)^2 + 3(10) + 8 & & & 7438 \\
& & & 6 & 6 \\
\hline
4(10)^4 + 4(10)^3 + 6(10)^2 + 2(10) + 8 & & & & 44628
\end{array}
$$

Now let us consider a more general product, say 7438 × 236.

7438 × 236 = 7438 × $[2(10)^2 + 3(10) + 6]$ which by the distributive property equals

$$(7438 \times 2)(10)^2 + (7438 \times 3)(10) + 7438 \times 6.$$

Thus we have reduced the problem to the sum of three **partial products**, each of which is a product with a one digit factor times a power of ten. The powers of ten, as we have seen, merely put an appropriate number of zeros on the end of each of the products with one digit factors. These partial products are then calculated one at a time from right to left and added in a column as follows.

$$
\begin{array}{r}
7438 \\
236 \\
\hline
\end{array}
$$

44628 $[= 7438 \times 6]$
223140 $[= (7438 \times 3)(10)]$
1487600 $[= (7438 \times 2)(10)^2]$

1755368

So 7438 × 236 = 1,755,368.

The zeros caused by the multiplications by 10 and 10^2 are usually left off so that the calculation looks like

7438
236

44628
22314
14876

1755368

Notice that the carried numbers are also left off, because there are too many of them. One just performs the carries mentally without writing them.

Notice that to multiply a number by 20, one merely multiplies it by 2 and puts a 0 on the end; to multiply a number by 300, one multiplies by 3 and puts two 0's on the end, etc.

An example in the base 5 system is $(234)_5 \times (312)_5$. The completed calculation with the ( )$_5$'s left off for convenience is

234
312

1023
234
1312

140113

Hence, $(234)_5 \times (312)_5 = (140113)_5$.

If we used the base 5 system commonly instead of the base 10 system, we would have memorized the following one-digit multiplication table for the base 5 system. For convenience, we again leave off the $( )_5$'s.

×	1	2	3	4
1	1	2	3	4
2	2	4	11	13
3	3	11	14	22
4	4	13	22	31

Of course, one should not try to memorize this table but just understand it.

Recall from Section 5.2 that in general, to perform one-digit calculations in a system of numerals with any base other than 10, it is usually easiest to do it using the word names for the numbers. For example to get $(2)_5 \times (4)_5$, we think "two times four is eight, and eight in the base five system is $(13)_5$." to get $(2)_3 + (2)_3$, we would think "two plus two is four, and four in the base three system is $(11)_3$."

Now let us look at $(212)_3 \times (122)_3$. The completed calculation with the $( )_3$'s left off is

$$
\begin{array}{r}
212 \\
122 \\
\hline
1201 \\
1201 \\
212 \\
\hline
112111
\end{array}
$$

Therefore, $(212)_3 \times (122)_3 = (112111)_3$.

Exercises

Find the following products by using the simplest written form of the calculation in each case.

1. 587×643;

2. 2506×108;

3. $(4342)_5 \times (342)_5$;

4. $(244)_5 \times (134)_5$;

5. $(21)_3 \times (121)_3$;

6. $(222)_3 \times (201)_3$.

7. Find the one-digit multiplication table for the base 4 system, and the base 12 system.

5.6   Division in our numeral system

It is to be emphasized that we are assuming that the reader already knows arithmetic including the processes of calculating used in addition, subtraction, multiplication, and division. We are not trying to show how to apply these familiar processes, but we are trying to explain why they are valid and just what is happening at each stage of a calculation.

We have already been exposed to the division algorithm, which states that if a, $b \in W$ and $b \neq 0$ then there exist unique whole numbers q and r such that

$$a = bq + r \text{ where } 0 \leq r < b.$$

Now $a \div b$ has a meaning in W if and only if $r = 0$. When $r = 0$, then $a \div b = q$. When $r \neq 0$, $a \div b$ has no meaning in W, but the quotient q and the remainder r still exist in W. The remainder r is what is left over when we attempt to divide a by b. The process of division is the process used to determine the quotient q and the remainder r.

Recall that bq is the largest multiple of b that is less than or equal to a. In order to find q, we keep subtracting multiples of b from a until what we have left is less than b but is greater than or equal to 0. Then q is the number of b's that we have subtracted and r is what we have left.

For example, let us try to divide 684 by 23. We will indicate the individual steps and explain what is happening each time. We begin by writing $684 \div 23$ in the form

$$23\overline{)684}$$

Next we subtract $23 \cdot 10 = 230$ from 684 as follows:

$$
\begin{array}{r}
23\overline{)684} \\
230 \\ \hline
454
\end{array} \quad 10
$$

Now, 454 is not less than 23, so we subtract more 23's say 10 of them to get:

$$
\begin{array}{r}
23\overline{)684} \\
230 \\ \hline
454 \\
230 \\ \hline
224
\end{array}
\quad
\begin{array}{l}
10 \\
\\
10
\end{array}
$$

Again, 224 is not less than 23, so we try to subtract say 9 more 23's, and obtain:

$$
\begin{array}{r|l}
23\overline{)684} & \\
230 & 10 \\
\hline
454 & \\
230 & 10 \\
\hline
224 & \\
207 & 9 \\
\hline
17 & \\
\end{array}
$$

What we have left is 17, which is less than 23, so we do not have to subtract any more multiples of 23. The number of 23's we have subtracted is $10 + 10 + 9 = 29$, so $684 = 23 \cdot 29 + 17$ where $0 \le 17 < 23$. The quotient is 29 and the remainder is 17, so $684 \div 23$ is not in W. We could have made better choices of the multiples of 23 we used, however, as follows:

$$
\begin{array}{r|l}
23\overline{)684} & \\
460 & 20 \\
\hline
224 & \\
207 & 9 \\
\hline
17 & \\
\end{array}
$$

With these excellent choices of the multiples of 23 to subtract, we could write the steps as follows (leaving off two of the zeros) in the normal form.

Step 1
$$
\begin{array}{r}
2 \\
23\overline{)684} \\
46 \\
\hline
224 \\
\end{array}
$$

Step 2
$$
\begin{array}{r}
29 \\
23\overline{)684} \\
46 \\
\hline
224 \\
207 \\
\hline
17 \\
\end{array}
$$

We will do one more example in a similar manner.

(1)
$$
\begin{array}{r}
1 \\
528\overline{)72864} \\
528 \\
\hline
2006 \\
\end{array}
$$

(2)
$$
\begin{array}{r}
13 \\
528\overline{)72864} \\
528 \\
\hline
2006 \\
1584 \\
\hline
4224 \\
\end{array}
$$

(3)
$$
\begin{array}{r}
138 \\
528\overline{)72864} \\
528 \\
\hline
2006 \\
1584 \\
\hline
4224 \\
4224 \\
\hline
0 \\
\end{array}
$$

So the quotient is 138 and the remainder is 0. We have

$$72864 = 528(138) + 0 \qquad 0 \le 0 < 528.$$

In this case, since the remainder is zero, $72864 \div 528 = 138$ and is in W.

Let us see how this works in the base 5 and base 3 systems. First consider $(332)_5 \div (23)_5$.

$$
\begin{array}{r}
23\overline{)332} \\
230 \quad 10 \\
\overline{102} \\
101 \quad 2 \\
\overline{1}
\end{array}
$$

Here the quotient is $(10)_5 + (2)_5 = (12)_5$ and the remainder is $(1)_5$, so $(332)_5 = (23)_5(12)_5 + (1)_5$, and hence $(332)_5 \div (23)_5$ is not in W.

As another example, take $(14143)_5 \div (32)_5$. Two ways of writing the calculation are:

$$
\begin{array}{rll}
 & & \quad\quad 234 \\
32\overline{)14143} & & \text{or} \quad 32\overline{)14143} \\
11400 \quad 200 & & \quad\quad 114 \\
\overline{2243} & & \quad\quad \overline{224} \\
2010 \quad 30 & & \quad\quad 201 \\
\overline{233} & & \quad\quad \overline{233} \\
233 \quad 4 & & \quad\quad 233 \\
\overline{0} & & \quad\quad \overline{0}
\end{array}
$$

Thus, $(14143)_5 = (234)_5(32)_5 + (0)_5$, so $(14143)_5 \div (32)_5 = (234)_5$ is in W.

Now we will take $(2002)_3 \div (21)_3$. We obtain

$$
\begin{array}{rll}
 & & \quad\quad 22 \\
21\overline{)2002} & & \text{or} \quad 21\overline{)2002} \\
1120 \quad 20 & & \quad\quad 112 \\
\overline{112} & & \quad\quad \overline{112} \\
112 \quad 2 & & \quad\quad 112 \\
\overline{0} & & \quad\quad \overline{0}
\end{array}
$$

Therefore, $(2002)_3 \div (21)_3 = (22)_3$.

Exercises

Perform each of the following divisions in both of the alternative ways for writing the calculations:

1. $72,072 \div 572$;   **2.** $437,531 \div 863$;

3. $(14443)_5 \div (23)_5$;   **4.** $(20101)_5 \div (42)_5$;

5. $(2121)_3 \div (12)_3$;   **6.** $(112111)_3 \div (212)_3$.

chapter 6

The integers

6.1 Introduction to integers

When we defined subtraction in W, we noticed that W is not closed under subtraction. (Recall that the definition of subtraction in W is that if a and b are in W, then $a - b$ is the whole number x such that $b + x = a$ if such a whole number x exists.) For example, $3 - 7$ does not exist in W because there is no whole number x such that $7 + x = 3$. In fact, we found that when a and b are in W, $a - b$ exists in W if and only if $a \geq b$. In order to remedy this situation, we shall develop the **system of integers**. Then there will be a number (not necessarily in W) for $a - b$ for every pair of whole numbers a and b. The system of integers is larger than the system of whole numbers and, in fact, contains the system of whole numbers.

Let us consider the Cartesian product, $I = W \times W$, of the set W of whole numbers with itself; that is, we consider

$$I = W \times W = \{(a, b) \mid a \in W \text{ and } b \in W\}.$$

We shall want (a, b) to turn out to be $a - b$, and our definitions of equality, addition, and multiplication in I will reflect this. We must be careful, however, to use only things that have been previously defined. Bearing this in mind, we shall continue.

Definition 6.1.1. (*Definition of equality in I*) *Two elements* (a, b) *and* (c, d) *of I are called* **equal**, *and we write* $(a, b) = (c, d)$, *if and only if* $a + d = b + c$ *in W.*

For example, $(3, 2)$, $(3, 5)$, $(0, 2)$, $(7, 6)$ are in I, and $(3, 5) = (0, 2)$ because $3 + 2 = 5 + 0$ in W. Similarly, $(3, 2) = (7, 6)$ because $3 + 6 = 2 + 7$.

Definition 6.1.2. (*Definition of addition and multiplication in* I) *Let* (a, b) *and* (c, d) *be any elements of* I, *then*

$$(a, b) + (c, d) = (a + c, b + d) \text{ and } (a, b)(c, d) = (ac + bd, ad + bc).$$

For example, $(2, 3) + (5, 2) = (2 + 5, 3 + 2) = (7, 5)$ and $(2, 3)(5, 2) = (2 \cdot 5 + 3 \cdot 2, 2 \cdot 2 + 3 \cdot 5) = (16, 19)$.

Definition 6.1.3. *The set* $I = W \times W$ *with the equality, addition, and multiplication of Definitions* 6.1.1 *and* 6.1.2 *is called the* **system of integers**.

From now on we shall also use the symbol I to denote the system of integers, not just the set $W \times W$. That is, we shall now denote by I the set $W \times W$ with equality, addition, and multiplication defined by Definitions 6.1.1 and 6.1.2. The individual elements (a, b) of I will be called **integers**.

In order to have the necessary freedom when using equalities of integers, we prove the next theorem indicating that equality of integers is an **equivalence relation**, that is, it satisfies the reflexive, symmetric, and transitive properties. This theorem is similar in nature to Theorems 3.1.1 and 3.8.1.

Theorem 6.1.1. *If* (a, b), (c, d), *and* (g, h) *are integers, then*

   (1) $(a, b) = (a, b)$ (*equality of integers is reflexive*);
   (2) *If* $(a, b) = (c, d)$ *then* $(c, d) = (a, b)$ (*equality of integers is symmetric*);
   (3) *If* $(a, b) = (c, d)$ *and* $(c, d) = (g, h)$ *then* $(a, b) = (g, h)$ (*equality of integers is transitive*).

   Proof.
   (1) $(a, b) = (a, b)$, because $a + b = b + a$ in W.
   (2) If $(a, b) = (c, d)$, then, by definition of equality, $a + d = b + c$, and hence $b + c = a + d$, because equality of whole numbers is symmetric. Therefore $c + b = d + a$ and hence $(c, d) = (a, b)$.
   (3) If $(a, b) = (c, d)$ and $(c, d) = (g, h)$, then $a + d = b + c$ and $c + h = d + g$. Hence

$$a + d + c + h = b + c + d + g,$$

which, by the cancellation property of addition in W, gives us

$$a + h = b + g.$$

Thus $(a, b) = (g, h)$.

The following theorem will help to simplify our development of the integers.

Theorem 6.1.2. *If a, b, and k are whole numbers, then*

(1) $(a + k, b + k) = (a, b)$;
(2) $(k, k) = (0, 0)$.

Proof. According to the definition of equality in I (Definition 6.1.1), $(a + k, b + k) = (a, b)$ because $a + k + b = b + k + a$ in W. Similarly, $(k, k) = (0, 0)$ because $k + 0 = k + 0$ in W.

Exercises

1. Why is it true that $(8, 2) = (6, 0)$ and $(3, 7) = (0, 4)$?
2. Is it true that $(2, 1) = (3, 5)$? Why?
3. Prove that if a, b, and h are whole numbers, then $(a + h, h) = (a, 0)$ and $(7 + h, 2 + h) = (5, 0)$.
4. Add:
   (a) $(2, 3) + (7, 1)$;   (b) $(0, 5) + (6, 3)$;   (c) $(3, 1) + (0, 0)$.
5. Multiply: (a) $(2, 3)(7, 1)$;   (b) $(0, 5)(6, 3)$;
   (c) $(3, 1)(0, 0)$;   (d) $(2, 0)(7, 1)$;   (e) $(0, 2)(0, 3)$;
   (f) $(5, 0)(2, 0)$;   (g) $(3, 0)(0, 2)$;   (h) $(0, 5)(6, 0)$.
6. By the definition of equality in I, $(4, 2) = (3, 1)$ and $(2, 5) = (6, 9)$.
   (a) Add $(4, 2) + (2, 5)$ and add $(3, 1) + (6, 9)$ and then see if your results are equal.
   (b) Multiply $(4, 2)(2, 5)$ and $(3, 1)(6, 9)$ and see if your results are equal.

6.2   Isomorphism

We shall see in this section that the system of integers is an enlargement of the system of whole numbers in that the latter system is essentially part of the system of integers.

For each integer (a, b), the numbers a and b are whole numbers and hence, by the trichotomy property of W, one and only one of the relations $a < b$, $a = b$, $a > b$ must hold. We will use this information to describe the following two kinds of integers:

(1) Those integers (a, b) for which $a \geq b$ in W.
(2) Those integers (a, b) for which $a < b$ in W.

Now $a \geq b$ if and only if there exists a whole number h (positive or zero) such that $a = b + h$. Then $(a, b) = (b + h, b + 0) = (h, 0)$ by Theorem 6.1.2. Similarly, $a < b$ if and only if there is a positive whole number k such that $a + k = b$. Then $(a, b) = (a + 0, a + k) = (0, k)$.

Thus there are two kinds of integers:

(1) Those that are equal to an integer of the form $(h, 0)$ where $h \geq 0$.

(2) Those that are equal to an integer of the form $(0, k)$ where $k > 0$.

For example, $(9, 7) = (7 + 2, 7 + 0) = (2, 0)$; $(2, 5) = (2 + 0, 2 + 3) = (0, 3)$; and $(3, 3) = (0, 0)$.

We shall see, however, that the set $W' = \{(h, 0) \mid h \in W\}$ of all integers of this first kind behaves the same under equality, addition, and multiplication as W does in the sense that we shall now demonstrate.

First, notice that if $h \in W$, the sets W and W' can be put in 1–1 correspondence using the pairing

$$0 \leftrightarrow (0, 0),\ 1 \leftrightarrow (1, 0),\ 2 \leftrightarrow (2, 0),\ \ldots,\ h \leftrightarrow (h, 0),\ \ldots.$$

For convenience of notation this 1–1 correspondence can be described by writing.

$$h \leftrightarrow h' = (h, 0) \qquad \text{for all } h \text{ of } W,$$

where the element of W' paired with h is $h' = (h, 0)$. Then if a and b are any two elements of W, they are paired with $a' = (a, 0)$ and $b' = (b, 0)$ respectively.

Evidently $a = b$ if and only if $a' = b'$; that is, $a = b$ if and only if $(a, 0) = (b, 0)$, by definition of equality in I.

If we obtain the sum of a and b, is this paired in the 1–1 correspondence with the sum of the things with which a and b are paired? In other words, is $(a + b)' = a' + b'$? Now $(a + b)' = (a + b, 0)$ and $a' + b' = (a, 0) + (b, 0)$. But $(a + b, 0) = (a, 0) + (b, 0)$ by definition of addition in I, so $(a + b)' = a' + b'$. That is, $(a + b)' = (a + b, 0) = (a, 0) + (b, 0) = a' + b'$. We say then that the 1–1 correspondence " preserves " addition.

Similarly, ab is paired with $(ab, 0)$ which is, in fact, equal to $(a, 0)(b, 0)$, the product of the things with which a and b are paired. That is, $(ab)' = (ab, 0) = (a, 0)(b, 0) = a'b'$. Consequently the 1–1 correspondence also " preserves " multiplication.

Thus W and W' behave the same under equality, addition, and multiplication and are essentially the same system, except for notation. Because of this, we shall identify W' with W and usually write a for $(a, 0)$ for all $a \in W$, and we shall also write $a = (a, 0)$. Thus we will consider that W is a subset of I. In particular, we will write $(7, 0) = 7$, $(1, 0) = 1$, and $(0, 0) = 0$.

The informal " preserves addition " is expressed more formally by " is isomorphic under addition," and " preserves multiplication " is more formally expressed by " is isomorphic under multiplication," according to the following definition.

Definition 6.2.1. *Let* $s \leftrightarrow s'$ *be a* 1–1 *correspondence between two sets S and S'* (*each with an equivalence relation* " $=$ " *defined*) *with the property that* $a = b$ *if and only if* $a' = b'$.

(1) *If S and S' both have an addition* " $+$ " *defined and* $(a + b)' = a' + b'$ *for all a, b of S, then S and S' are said to be **isomorphic under addition** and the* 1–1 *correspondence is called an **isomorphism under addition**.*

(2) *If S and S' both have a multiplication defined and* $(ab)' = a' b'$ *for all a, b of S, then S and S' are said to be **isomorphic under multiplication** and the* 1–1 *correspondence is called an **isomorphism under multiplication**.*

According to this definition, what we have proved above about the behavior of W and W' can be stated as follows.

Theorem 6.2.1. *The* 1–1 *correspondence* $h \leftrightarrow h' = (h, 0)$ *of W and* $W' = \{(h, 0) \mid h \in W\}$ *is an isomorphism under addition and multiplication.*

It is to be emphasized that because we have decided to identify W' with W, W will be considered to be a subset of I (because $W' \subset I$).

To illustrate further the idea of isomorphism, let us consider the following sets:

$$S = \{1, 2, 3, \ldots\} \quad \text{and} \quad S' = \{2, 4, 6, \ldots\}.$$

Of course, S is the set of positive whole numbers and S' is the set of positive even whole numbers. As we have seen in Section 3.8, the pairing $n \leftrightarrow n' = 2n$, for all $n \in S$, is a 1–1 correspondence of S and S'. Is it an isomorphism under addition and multiplication?

Let a and b be any two elements of S; then a is paired with $a' = 2a$ and b is paired with $b' = 2b$. If $a = b$ then $2a = 2b$ and hence $a' = b'$. Conversely if $a' = b'$ then $2a = 2b$, so $a = b$. Thus $a = b$ if and only if $a' = b'$.

To see whether the 1–1 correspondence is an isomorphism under addition, we must see whether $(a + b)' = a' + b'$ for all a, b of S. Now

$$(a + b)' = 2(a + b) \quad \text{and} \quad a' + b' = 2a + 2b.$$

But $2(a + b) = 2a + 2b$ by the distributive property. Therefore, $(a + b)' = a' + b'$, and we have an isomorphism under addition.

To see if we have an isomorphism under multiplication, we must see whether $(ab)' = a' b'$ for all a, b of S. Now

$$(ab)' = 2(ab) \quad \text{and} \quad a' b' = (2a)(2b).$$

But $2(ab) = 2ab$ and $(2a)(2b) = 4ab$, so $(ab)' \neq a' b'$ and we do not have an isomorphism under multiplication.

It is important to understand the difference in the meaning of $(a + b)'$ and $a' + b'$, and the difference in the meaning of $a'b'$ and $(ab)'$. To obtain $(a + b)'$, first add a to b to get $a + b$, and then find the thing paired off with the number that is the sum $a + b$. To obtain $a' + b'$ first find a' (the thing paired off with a) and b' (the thing paired off with b) and then add a' to b' to get $a' + b'$. Similarly, to determine $(ab)'$ first multiply a by b to get ab, and then find the thing paired off with the number that is ab. To obtain $a'b'$, first find a' (the thing paired off with a) and b' (the thing paired off with b) and multiply a' by b' to get $a'b'$.

Although $(a + b)'$ and $a' + b'$ have different meanings, they *may* turn out to equal the same thing and hence be equal to each other. Similarly, even though $(ab)'$ and $a'b'$ have different meanings, they *may* turn out to equal the same thing and hence be equal to each other.

Exercises

Let S be the set of all positive whole numbers. In each of the following exercises, determine whether the correspondence $n \leftrightarrow n'$ given for each S' is an isomorphism under (a) addition, (b) multiplication.

1. $n \leftrightarrow n' = 3n$ for all n in S, where $S' = \{3n \mid n \in S\}$;

2. $n \leftrightarrow n' = n + 1$ for all n in S, where $S' = \{n + 1 \mid n \in S\}$;

3. $n \leftrightarrow n' = (0, n)$ for all n in S, where $S' = \{(0, n) \in I \mid n \in S\}$;

4. $n \leftrightarrow n' = 2n + 1$ for all n in S, where $S' = \{2n + 1 \mid n \in S\}$;

5. $n \leftrightarrow n' = 2^n$ for all n in S, where $S' = \{2^n \mid n \in S\}$.

6.3   Addition, multiplication, subtraction, and division of integers

Before seeing why (a, b) actually does turn out to be $a - b$, we shall obtain some properties of the integers. First we will show that addition and multiplication of integers are not ambiguous, that is, they are **well defined**. A similar thing for whole numbers was done in Theorems 4.2.1 and 4.2.2.

Theorem 6.3.1. *If (a, b), (a', b'), (c, d), (c', d') are integers such that*

$$(a, b) = (a', b') \qquad and \qquad (c, d) = (c', d'),$$

then

(1) $(a + c, b + d) = (a' + c', b' + d')$;

(2) $(ac + bd, ad + bc) = (a'c' + b'd', a'd' + b'c')$.

Proof. By definition of equality of integers, we are given

$$\text{(a) } a + b' = b + a' \qquad \text{and} \qquad \text{(b) } c + d' = d + c',$$

and we must prove each of the following:

(1)′ $\qquad\qquad a + c + b' + d' = b + d + a' + c',$

(2)′ $\qquad\qquad ac + bd + a'd' + b'c' = ad + bc + a'c' + b'd'.$

Adding equations (a) and (b), we get

$$a + b' + c + d' = b + a' + d + c',$$

which is essentially (1)′.

To obtain (2)′, we multiply equation (a) by c and then d to get

(c) $\qquad\qquad\qquad ac + b'c = bc + a'c,$

(d) $\qquad\qquad\qquad bd + a'd = ad + b'd,$

and we multiply equation (b) by a' and b' giving

(e) $\qquad\qquad\qquad a'c + a'd' = a'd + a'c',$

(f) $\qquad\qquad\qquad b'd + b'c' = b'c + b'd'.$

Now by adding equations (c), (d), (e), and (f), we get

$$ac + b'c + bd + a'd + a'c + a'd' + b'd + b'c' =$$
$$bc + a'c + ad + b'd + a'd + a'c' + b'c + b'd'.$$

If we cancel the terms which appear on both sides of this equation, we obtain (2)′.

In less rigorous language, Theorem 6.3.1 says that when equal integers are added to equal integers the results are equal, and that when equal integers are multiplied by equal integers the results are equal.

We shall define subtraction and division in I similarly to the way we defined these operations in W.

Definition 6.3.1. *If u and v are integers, then u − v is the number x such that*

$$v + x = u.$$

Definition 6.3.2. *If u and v are integers, then u ÷ v is the number x such that*

$$vx = u.$$

That is, *u − v is the number which when added to v produces u*, and *u ÷ v is the number which when multiplied by v produces u.*

Notice, similarly to the corresponding statements for W, it follows direct- ly from the definitions of subtraction and division in I (Definitions 6.3.1 and 6.3.2) that $v + (u - v) = u$, $(u + w) - w = u$, $v(u \div v) = u$, and $(uw) \div w = u$, when these indicated operations exist, These facts will be used later.

Definitions 6.3.1 and 6.3.2 do not, themselves, show the existence of $u - v$ and $u \div v$. In fact, $u \div v$ only exists in I in certain instances. In particular $2 \div 3$ does not exist in I because there is no integer x such that $3x = 2$. But $6 \div 3$ does exist in I because the integer 2 is such that $3 \cdot 2 = 6$. However, we shall see in Theorem 6.3.4 that $u - v$ always exists in I; that is, we shall see that I is closed under subtraction.

When $u \div v$ does exist in I it is unique, as we shall see in Section 6.7. We now arrive at a theorem that has many uses. In particular, it can be used to see that when $u - v$ exists in I it is unique.

Theorem 6.3.2. *Let u, v, and w be integers. Then $u = v$ if and only if $u + w = v + w$.*

Proof. Since equality is reflexive in I, $w = w$. If $u = v$, we can use the fact that addition is not ambiguous in I (Theorem 6.3.1) to get $u + w = v + w$. To prove conversely that if $u + w = v + w$ then $u = v$, we let $u = (a, b)$, $v = (c, d)$, and $w = (e, f)$ where a, b, c, d, e, f are in W. Then, if $u + w = v + w$, we have

$$(a + e, b + f) = (c + e, d + f)$$

by definition of addition, and hence by definition of equality

$$a + e + d + f = b + f + c + e.$$

Therefore, by the cancellation property of addition in W (Theorem 4.7.5), we can cancel e and f to obtain $a + d + b + c$. Hence, by definition of equality in I, $u = v$.

Note that the "if" part of the double implication of Theorem 6.3.2 is the cancellation property of addition for I. We will use it to prove the next theorem.

Theorem 6.3.3. *When $u - v$ exists in I, it is unique.*

Proof. If x and x' are numbers such that $v + x = u$ and $v + x' = u$, then $v + x = v + x'$. Therefore, by Theorem 6.3.2, $x = x'$, and hence if $u - v$ exists, it is unique.

In the next section we shall arrive at the familiar notation for all integers. We then shall discuss the familiar ways of actually adding, subtracting, multi- plying, and dividing integers. We now have a theorem indicating that the integers have all of the familiar properties of Theorem 4.3.2 satisfied by W and are closed under subtraction.

Theorem 6.3.4. *If u, v, and w are in I, then*

(1) $u + v$ is in I;   (2) $u + v = v + u$;
(3) $(u + v) + w = u + (v + w)$;   (4) $u + 0 = u$;
(5) $u - v$ is in I;   (6) uv is in I;
(7) $uv = vu$;   (8) $(uv)w = u(vw)$;
(9) $u \cdot 1 = u$;   (10) $u(v + w) = uv + uw$,
$$(v + w)u = vu + wu.$$

Proof. We shall prove parts (1), (2), (3), (4), and (7), leaving the proofs of parts (6), (8), (9), and (10) as exercises. Part (5) will be proved in Section 6.6, when we shall be better equipped to deal with it. Notice that each of the equalities of this theorem can be proved by first using the definitions of addition and multiplication to obtain the left-hand side of the equation and then using these definitions to obtain the right-hand side of the equation. Each proof is then completed by noting that the left side and right side (using the properties of W) turn out to be the same thing. The transitive property of equality is then used to see that they are equal.

Let $u = (a, b)$, $v = (c, d)$, and $g = (e, f)$, where a, b, c, d, e, and f are whole numbers.

(1) $u + v = (a, b) + (c, d) = (a + c, b + d)$, by definition of addition in I. But by the closure property of addition in W, $a + c$ and $b + d$ are whole numbers, and therefore $(a + c, b + d)$ is in I.

(2) By definition of addition in I, and the commutative property of addition in W, we have

$$u + v = (a, b) + (c, d) = (a + c, b + d)$$

and

$$v + u = (c, d) + (a, b) = (c + a, d + b) = (a + c, b + d).$$

Therefore $u + v = v + u$.

(3) By definition of addition in I and the associative property of addition in W.

$$(u + v) + w = ((a, b) + (c, d)) + (e, f) = (a + c, b + d) + (e, f)$$
$$= (a + c + e, b + d + f).$$

and

$$u + (v + w) = (a, b) + ((c, d) + (e, f)) = (a, b) + (c + e, d + f)$$
$$= (a + c + e, b + d + f).$$

Consequently, $(u + v) + w = u + (v + w)$.

(4) Recall from Section 6.2 that $0 = (0, 0)$. Then by definition of addition, $u + 0 = (a, b) + (0, 0) = (a + 0, b + 0) = (a, b) = u$. Hence $u + 0 = u$.

(7) By definition of multiplication and the commutative properties of addition and multiplication in W,

$$uv = (a, b)(c, d) = (ac + bd, ad + bc)$$

and

$$vu = (c, d)(a, b) = (ca + db, cb + da) = (ac + bd, ad + bc).$$

Therefore $uv = vu$.

We next obtain the cancellation property for subtraction in I.

Theorem 6.3.5. *Let u, v, and w be integers. Then $u = v$ if and only if $u - w = v - w$.*

Proof. Let $u = v$. To show that this implies that $u - w = v - w$, we show that $v - w$ is the number which when added to w produces u, and hence by definition of subtraction is equal to $u - w$. Now $w + (v - w) = v$ by definition of subtraction. But also $u = v$, so $w + (v - w) = u$, and hence $u - w = v - w$.

Conversely if $u - w = v - w$, then, by Theorem 6.3.2, $(u - w) + w = (v - w) + w$. But $(u - w) + w = u$ and $(v - w) + w = v$, and therefore $u = v$.

Exercises

1. Rewrite Theorem 6.3.4 indicating the word description of each property. (See Theorem 4.3.2.)

2–5. Prove parts (6), (8), (9), and (10) of Theorem 6.3.4.

6.4 Positive and negative integers, the negative of an integer

We shall define positive and negative integers as follows:

Definition 6.4.1. *An integer that is equal to an integer of the form $(h, 0)$ where $h > 0$ is called a **positive** integer, and an integer equal to one of the form $(0, k)$ where $k > 0$ is called a **negative** integer.*

Since $(h, 0) = h$, the positive integers are the positive whole numbers; that is, they are the numbers 1, 2, 3, 4, ... Before arriving at the familiar notation for the negative integers, we shall consider what is meant by the negative of an integer.

Definition 6.4.2. *If u is an integer, then the number x such that $u + x = 0$ is called the **negative** of u and is written as* $-u$.

Notice that by the definition of subtraction, the definition of $-u$ could have been expressed by saying that $-u$ is defined to be $0 - u$, because $0 - u$ means the number x such that $u + x = 0$.

For example, $(2, 5) + (5, 2) = (7, 7) = (0, 0) = 0$, so $(5, 2)$ is the negative of $(2, 5)$ and we write $(5, 2) = -(2, 5)$. It is to be noticed that since addition is commutative in I, $(5, 2) + (2, 5) = 0$. Consequently, we also have $(2, 5) = -(5, 2)$. In general, the negative of any integer is equal to the integer obtained when the components of the ordered pair are interchanged, as the following theorem indicates:

Theorem 6.4.1. *The negative of the integer (a, b) is (b, a); that is, $(b, a) = -(a, b)$.*

Proof.

$$(a, b) + (b, a) = (a + b, b + a) = (a + b, a + b) = (0, 0) = 0,$$

therefore $(b, a) = -(a, b)$.

Notice that this theorem also shows that *the negative of every integer exists and is also an integer*.

We can now arrive at the familiar, convenient notation for the negative integers. By Theorem 6.4.1, $(0, k) = -(k, 0) = -k$, because we are writing k for $(k, 0)$. Thus every integer of the form $(0, k)$ can be expressed in the form $-k$, where k is a whole number, and hence every integer is either a whole number (positive or zero) or the negative of a whole number.

For example, $(5, 2) = (3 + 2, 0 + 2) = (3, 0) = 3$, and $(2, 5) = (0 + 2, 3 + 2) = (0, 3) = -3$. Also $0, 7, -2, 5, -7$, and -6 are integers, and every integer can be similarly expressed as a positive integer, a negative integer, or zero, and we have the following theorem.

Theorem 6.4.2. *If u is an integer then one and only one of the following holds:*
    (1) *u is positive*   (2) $u = 0$ (3) *u is negative.*

Proof. We have seen in Section 6.2 that every integer is equal to one of the form $(h, 0)$ where $h \geq 0$ or one of the form $(0, k)$ where $k > 0$. When $h = 0$, $(h, 0) = (0, 0) = 0$, and when $h > 0$, $(h, 0)$ is positive by definition. Similarly, by definition, $(0, k)$ is negative. Thus at least one of the three possibilities stated in the theorem holds. Also, an integer cannot be both positive and equal to zero, for if $(h, 0) = (0, 0)$, then, by definition of equality in I, $h = 0$, and we cannot have $h > 0$. Similarly u cannot be negative and equal to zero. Finally, u cannot be both a positive integer and a negative integer, for if $(h, 0) = (0, k)$, then $h = 0$ and $k = 0$ by definition of equality in I, and we cannot have $h > 0$ or $k > 0$.

It is important to realize that the negative of an integer is not necessarily a negative integer. If u is an integer, then $-u$ might be a positive integer, zero, or a negative integer, depending on what u is. For example, the negative of 5 is -5, a negative integer; but what is the negative of -2? Well, -2 is the integer $(0, 2)$, and, according to Theorem 6.4.1, $-(0, 2) = (2, 0)$, which is 2. Consequently the negative of -2 is 2; that is, $-(-2) = 2$. In this case the negative of the integer -2 is the positive integer 2.

More generally, we have the following theorem:

Theorem 6.4.3. *Let u be an integer, then*
(1) *If u is a positive integer, then $-u$ is a negative integer;*
(2) *If $u = 0$ then $-u = 0$;*
(3) *If u is a negative integer, then $-u$ is a positive integer.*

Proof.
(1) If u is a positive integer, then, by definition, $u = (h, 0)$ where $h > 0$. But by Theorem 6.4.1, $-u = (0, h)$; that is, $-u$ is a negative integer.
(2) If $u = 0$ then $u = (0, 0)$ and by Theorem 6.4.1, $-u = (0, 0) = 0$.
(3) If u is a negative integer, then, by definition, $u = (0, k)$ where $k > 0$. Therefore, by Theorem 6.4.1, $-u = (k, 0)$, a positive integer.

We will now be able to see that (a, b) does equal $a - b$, as we had originally planned with our definitions of equality, addition, and multiplication.

Theorem 6.4.4. *The integer $(a, b) = a - b$.*

Proof. By using the fact that $a = (a, 0)$, $b = (b, 0)$, and the definitions of addition and equality in I, we have

$$b + (a, b) = (b, 0) + (a, b) = (b + a, b + 0) = (a, 0) = a.$$

Therefore, by definition of subtraction, $(a, b) = a - b$.

For example: $(5, 2) = 5 - 2 = 3$; $(2, 0) = 2 - 0 = 2$; $(2, 5) = (2, 2 + 3) = (0, 3) = -3 = 2 - 5$; and $8 - 10 = (8, 10) = (8, 8 + 2) = (0, 2) = -2$. Thus $2 - 5 = -3$ and $8 - 10 = -2$.

We conclude this section with a theorem that is often useful.

Theorem 6.4.5. *If u is an integer, then $(-1)u = -u$.*

Proof. Let $u = (a, b)$ where a and b are in W. Then by definition of multiplication in I and Theorem 6.4.1,
$(-1)u = (0, 1)(a, b) = (0 \cdot a + 1 \cdot b, 0 \cdot b + 1 \cdot a) = (b, a) = -(a, b) = -u$.
Thus $(-1)u = -u$.

Exercises

1. What is the negative of each of the following integers?
   (a) 3;   (b) 17;   (c) 0;   (d) −7;   (e) −86;   (f) −15.
2. Express each of the following integers in the familiar form as a whole number or the negative of a whole number, in which parentheses do not appear.
   (a) (7, 2);   (b) (9, 1);   (c) (3, 3);   (d) (3, 5);   (e) (2, 7);
   (f) (4, 71).
3. Find the following differences: (Note how we saw above that $2 - 5 = -3$ and $8 - 10 = -2$.)
   (a) $7 - 2$;   (b) $3 - 5$;   (c) $2 - 7$;   (d) $4 - 71$;   (e) $62 - 87$;
   (f) $0 - 8$.
4. Use the *definition* of the negative of an integer to prove that if u is an integer, then $-(-u) = u$.
5. Use Theorem 6.4.1 to prove that if $u = (a, b)$ is an integer, then $-(-u) = u$.

6.5   Integers on the number line

The integers consist of the whole numbers and the negatives of the whole numbers, and in order to put them on the number line, we shall thus place the negative integers on the number line on which the whole numbers have already been placed. We have seen that a negative integer is the negative, $-a$, of a positive integer a. To determine the point labeled $-a$, we take the point that is the same distance to the left of zero as a is to the right of zero.

For example, the negative integers -1, -2, -3 appear on the line as follows:

Since, for a positive integer a, the integers a and $-a$ are each the negative of the other and are the same distance from the origin in opposite directions, it is often convenient to consider them as "opposites." Thus we may consider 5 and -5 as "opposites." With this thought in mind, let us now try to use the number line to see how we might expect combinations of positive and negative integers to behave under the operations of addition, subtraction, multiplication, and division.

It is to be emphasized that we are now speculating about what we might expect and are not proving anything. In order to prove that addition, subtraction, multiplication, and division of positive and negative integers behave in a certain way, we shall have to go back to the appropriate definitions and theorems and construct proofs.

We shall try to add, subtract, multiply, and divide integers u and v on the number line in the same sort of way that we used for whole numbers in Chapter 4. Thus we will let u and v be integers and try to proceed as follows:

(a) *To add $u + v$, go v units to the right of u.*

(b) *To subtract $u - v$, go v units to the left of u.*

(c) *To multiply uv, go u steps, of v units each, to the right of the origin.*

(d) *To divide $u \div v$, determine how many v-unit steps to the right are needed to go from the origin to u.*

When neither u nor v is negative, we can proceed as before. But, for example, what is meant by "go -2 units to one side of u"? Considering -2 and 2 as "opposites," it means to go 2 units to the *opposite* side of u. Similarly, to go u steps of -2 units to the *right* of the origin means to go u steps of 2 units to the *left* of the origin. Also, to go -5 steps of v units to the *right* of the origin means to go 5 steps of v units to the *left* of the origin, and to determine how many -2-unit steps to the *right* are needed to go from the origin to u, we determine how many 2-unit steps to the *left* are needed to go from the origin to u.

Let us consider the following more specific examples:

(1) To add $6 + (-2)$, go 2 units to the left of 6 to arrive at 4. Thus $6 + (-2) = 4$.

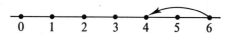

(2) To subtract $5 - (-2)$, go 2 units to the right of 5 to arrive at 7. Thus $5 - (-2) = 7$.

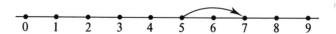

(3) To multiply $5(-2)$, go 5 steps of 2 units to the left of the origin, to arrive at -10. Thus $5(-2) = -10$.

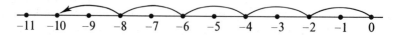

(4) To divide $6 \div (-2)$, determine how many 2-units steps to the left are needed to go from the origin to 6. Since it takes three 2-unit steps to the *right* to go from the origin to 6, it takes minus three 2-unit steps to the *left* to go from the origin to 6. Thus $6 \div (-2) = -3$.

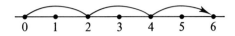

(5) To add $(-5) + 2$, go 2 units to the right of -5 to arrive at -3. Thus $(-5) + 2 = -3$.

(6) To subtract $(-5) - 2$, go 2 units to the left of -5 to arrive at -7. Thus $(-5) - 2 = -7$.

(7) To multiply $(-5)(2)$, go 5 steps of 2 units to the left of the origin to arrive at -10. Thus $(-5)(2) = -10$.

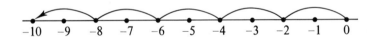

(8) To divide $(-6) \div 3$, determine how many 3-unit steps to the right are needed to go from the origin to -6. Since it takes two 3-unit steps to the *left* to go from the origin to -6, it takes minus two 3-units steps to the *right* to go from the origin to -6. Thus $(-6) \div 3 = -2$.

(9) To add $(-5) + (-2)$, go 2 units to the left of -5 to arrive at -7. Thus $(-5) + (-2) = -7$.

(10) To subtract $(-5) - (-2)$, go 2 units to the right of -5 to arrive at -3. Thus $(-5) - (-2) = -3$.

(11) To multiply $(-5)(-2)$, go -5 steps of 2 units each to the left of the origin. That is, go 5 steps of 2 units to the right of the origin to arrive at 10. Thus $(-5)(-2) = 10$.

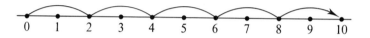

(12) To divide $(-6) \div (-2)$, determine how many 2-unit steps to the left are needed to go from the origin to -6. We see that three such steps are needed, so $(-6) \div (-2) = 3$.

Note that in examples (3), (4), (7), and (8) where only one of the numbers in the product or quotient is negative, the result is negative, but in examples (11) and (12) where both numbers in the product or quotient are negative, the result is positive. We shall see in the next section that positive and negative integers always combine this way, and that the results obtained in examples (1) to (12) are correct.

Exercises

1. Draw a number line with 1-inch units, and locate each of the following points on it: 0, 1, 2, 3, 4, -1, -2, -3, -4.

2. Find each of the following mechanically on the number line, and draw appropriate figures:

    (a) $(-7) - 2$;   (b) $7 - (-2)$;   (c) $3(-2)$;   (d) $6 \div (-2)$;

    (e) $(-2)(-3)$;   (f) $(-5) - (-3)$;   (g) $(-8) \div (-2)$;

    (h) $8 + (-3)$;   (i) $(-4)(-3)$;   (j) $(-3)(-5)$;

    (k) $(-12) \div (-3)$;   (l) $(-10) \div 2$;   (m) $(-3) + (-4)$.

6.6 Calculations with integers

Now let us see how to actually add, subtract, multiply, and divide integers using the familiar notation for them as either whole numbers or the negatives of whole numbers. Because we already know how to add, multiply, and divide whole numbers, we only need to see how to perform these operations when one or both of the given numbers is a negative integer. In the case of subtraction, we know how to find $u - v$ when u and v are whole numbers and $u \geq v$, but not when $u < v$. When $u < v$, we use part (1) of the next theorem to express $u - v$ as $-(v - u)$. For example, $1 - 3 = -(3 - 1) = -2$. To perform a subtraction where one or both of the given numbers is a negative integer, we use (2), (3), or (4) of the next theorem.

By appropriate use of the parts of the following theorem, we can add, subtract, multiply, and divide in the system of integers and obtain the results indicated by our mechanical process on the number line.

Theorem 6.6.1. *Let a and b be any whole numbers. Then*
   (1) $(-b) + a = a + (-b) = a - b = -(b - a)$;
   (2) $(-a) + (-b) = (-a) - b = (-b) - a = -(a + b)$;
   (3) $(-a) - (-b) = (-a) + b = b - a = -(a - b)$;
   (4) $a - (-b) = a + b$;
   (5) $(-a)b = a(-b) = -(ab)$;
   (6) $(-a)(-b) = ab$;
   (7) If $b \neq 0$, $(-a) \div b = a \div (-b) = -(a \div b)$;
   (8) If $b \neq 0$, $(-a) \div (-b) = a \div b$.

Proof.

(1) By the commutative property in I, our notation for integers, and Theorem 6.4.4, we have

$$(-b) + a = a + (-b) = (a, 0) + (0, b) = (a, b) = a - b.$$

Also by Theorems 6.4.1 and 6.4.4,

$$a - b = (a, b) = -(b, a) = -(b - a).$$

Therefore,

$$(-b) + a = a + (-b) = a - b = -(b - a).$$

(2) By our notation and the definition of addition in I,

$$(-a) + (-b) = (0, a) + (0, b) = (0, a + b) = -(a + b).$$

Also, when b is added to $(-a) + (-b)$, we get $(-a) + (-b) + b = -a + 0 = -a$. Consequently, by definition of subtraction, $(-a) - b = (-a) + (-b)$. Similarly, $(-b) - a = (-b) + (-a) = (-a) + (-b)$, and therefore

$$(-a) + (-b) = (-a) - b = (-b) - a = -(a + b).$$

(3) We will first use the definition of subtraction to show that $(-a) - (-b) = b - a$. Now

$$(-b) + (b - a) = (0, b) + (b, a) = (0 + b, a + b) = (0, a) = -a$$

(reasons for steps to be supplied by the reader). Therefore, by the definition of subtraction, $(-a) - (-b) = b - a$. The proof that $(-a) + b = b - a = -(a - b)$ is similar to the proof of (1) and is left as an exercise.

(4) $(-b) + (a + b) = (0, b) + (a + b, 0) = (a + b, b + 0) = (a, 0) = a$. Hence, by the definition of subtraction, $a - (-b) = a + b$.

(5) $(-a)b = (0, a)(b, 0) = (0 \cdot b + a \cdot 0, 0 \cdot 0 + a \cdot b) = (0, ab) = -(ab)$.
   $a(-b) = (a, 0)(0, b) = (a \cdot 0 + 0 \cdot b, ab + 0 \cdot 0) = (0, ab) = -(ab)$.
Therefore, $(-a)b = a(-b) = -(ab)$.

(6) $(-a)(-b) = (0, a)(0, b) = (0 \cdot 0 + ab, 0 \cdot b + a \cdot 0) = (ab, 0) = ab$.

(7) By definition of division, $b(a \div b) = a$, and we use this together with the multiplication properties of negatives of parts (5) and (6) to get: $b[-(a \div b)] = -[b(a \div b)] = -a$, hence, by definition of division, $(-a) \div b = -(a \div b)$. Similarly, $(-b)[-(a \div b)] = b(a \div b) = a$, so $a \div (-b) = -(a \div b)$. Therefore $(-a) \div b = a \div (-b) = -(a \div b)$.

(8) Similarly to (7), $(-b)(a \div b) = -[b(a \div b)] = -a$, so $(-a) \div (-b) = a \div b$.

Theorem 6.6.1 allows us to write every addition, multiplication, subtraction, or division in I in terms of a similar operation in W. Hence we may now do each of these operations in I. Note that it also follows from Theorem 6.6.1 that for any integers (not just whole numbers) u and v,

$$u - v = -(v - u).$$

Examples:

(a) $-7 + 3 = 3 - 7 = -(7 - 3) = -4$.
(b) $-11 + 6 = -(11 - 6) = -5$.
(c) $(-3) - 2 = -(3 + 2) = -5$: $\quad (-3) + (-2) = -(3 + 2) = -5$.
(d) $(-2) - (-3) = -2 + 3 = 3 - 2 = 1$.
(e) $(-7) - (-2) = -7 + 2 = -(7 - 2) = -5$.
(f) $8 - (-2) = 8 + 2 = 10$. $\qquad$ (g) $(-4)(3) = -(4 \cdot 3) = -12$.
(h) $2(-6) = -(2 \cdot 6) = -12$. $\qquad$ (i) $(-7)(-4) = 7(4) = 28$.
(j) $(-12) \div (-3) = 12 \div 3 = 4$. $\qquad$ (k) $(-15) \div (3) = -(15 \div 3) = -5$.

It is now convenient to complete the proof of part (5) of Theorem 6.3.4, which states that I is closed under subtraction.

Proof of Theorem 6.3.4, *part* (5). If u and v are whole numbers such that $u \geq v$, then, by definition of $\geq$, there is a whole number h such that $v + h = u$. But by definition of subtraction, $h = u - v$, so $u - v$ is a whole number and is therefore in I. If u and v are whole numbers and $u < v$, then by Theorem 6.6.1, $u - v = -(v - u)$. So $u - v$ is the negative of the integer $v - u$ and hence is in I. Thus we have proved closure when u and v are whole numbers.

If either one or both of u and v are negative integers, then their difference can be expressed by parts (1), (2), (3), or (4) of Theorem 6.6.1 as a difference of whole numbers, or the negative of such a difference, and hence is in I by the preceding paragraph. Thus, I is closed under subtraction.

Exercises

1. If u and v are integers, complete each of the following:
    (a) $-u$ is negative if and only if u is _____;
    (b) $-u$ is zero if and only if u is _____;
    (c) $-u$ is positive if and only if u is _____;
    (d) $u - v$ is positive if and only if $v - u$ is _____;
    (e) $u - v$ is negative if and only if $v - u$ is _____;
    (f) $u - v$ is zero if and only if $v - u$ is _____.
2. Finish the proof of part (3) of Theorem 6.6.1; that is, prove that $(-a) + b = b - a = -(a - b)$, when a and b are whole numbers.

Perform the indicated operations:

3. $-7 + 1$;   4. $-(-6)$;
5. $-9 - 7$;   6. $(-7) + (-6)$;
7. $(-5) - (-9)$;   8. $8 + 7 - 22$;
9. $(-21) \div (3)$;   10. $(-2)(-3)$;
11. $(-9)(6)$;   12. $-7(6 - 8)$;
13. $(-3)(-4) + 8(-6) - 7(9) - 8$;   14. $8(-11)$;
15. $(36) \div (-12)$;   16. $(-48) \div (-4)$.

6.7 Additional properties of addition, subtraction, multiplication, and division

In this section, we shall obtain some additional properties of the integers under the four fundamental operations. These properties are useful in many situations. The first theorem of this section helps greatly in simplifying many expressions involving subtraction. The proof is based primarily on the definition of subtraction and not on the special way that integers add and subtract. For this reason, the same proof can be used to prove the corresponding result for W or for the rational numbers or real numbers to be defined in Chapters 8 and 9.

Theorem 6.7.1. *If u, v, and w are integers, then*
   (1) $(u + v) - w = u + (v - w) = (u - w) + v$;
   (2) $(u - v) - w = (u - w) - v$.

Proof of (1). By definition of subtraction, $(u + v) - w$ is the number which when added to w will equal $u + v$. We shall show that each of the numbers $u + (v - w)$ and $(u - w) + v$ has this property and hence is equal to $(u + v) - w$.

Since, by definition of subtraction, $(v - w) + w = v$,

$$u + (v - w) + w = u + [(v - w) + w] = u + v.$$

Similarly, since $(u - w) + w = u$,

$$(u - w) + v + w = (u - w) + w + v = u + v.$$

Therefore,

$$u + (v - w) = (u + v) - w \quad \text{and} \quad (u - w) + v = (u + v) - w,$$

and (1) is proved.

To prove (2), we must show, by definition of subtraction, that when $(u - w) - v$ is added to w we get $u - v$. To do this we use part (1) of the theorem to obtain

$$[(u - w) - v] + w = [(u - w) + w] - v.$$

But $(u - w) + w = u$, so $[(u - w) - v] + w = u - v$ and hence, by definition of subtraction, $(u - v) - w = (u - w) - v$.

It follows from (1) that we may subtract an integer from a sum of any number of integers by subtracting it from any one of the summands. For example, if u, v, r, s, w are integers,

$$(u + v + r + s) - w = u + (v - w) + r + s = u + v + (r - w) + s$$
$$= (u - w) + v + r + s = u + v + r + (s - w)$$

Other examples are:

   (1) $(6x + 7) - 2x = (6x - 2x) + 7 = 4x + 7,$
   (2) $(6x + 7) - 2 = 6x + (7 - 2) = 6x + 5,$

(3) $8x + 2y + z - 3x = 8x - 3x + 2y + z = 5x + 2y + z,$
(4) $(9x - 3) - 2x = (9x - 2x) - 3 = 7x - 3.$

Theorem 6.7.2. *Let u, v, and w be integers. Then*

(1) $u \cdot 0 = 0 \cdot u = 0;$
(2) $uv = 0$ if and only if $u = 0$ or $v = 0$;
(3) $u(v - w) = uv - uw$ and $(v - w)u = vu - wu.$

Proof.
(1) Let $u = (a, b)$ where a and b are in W. Then

$$0 \cdot u = u \cdot 0 = (a, b)(0, 0) = (a \cdot 0 + b \cdot 0, a \cdot 0 + b \cdot 0) = (0, 0) = 0.$$

(2) If $u = 0$ or $v = 0$, then $uv = 0$ by (1). Conversely, we shall prove that if $u \neq 0$ and $v \neq 0$, then $uv \neq 0$. Now by Theorem 6.4.2, an integer is not zero if and only if it is positive or negative. But by Theorem 4.3.1, the positive integers (which are the same as the positive whole numbers) are closed under multiplication, and hence when u and v are both positive, their product uv is positive. Thus it follows from Theorem 6.6.1, part (6), that if u and v are either both positive or both negative, uv is positive and hence not zero. Similarly, by part (5) of Theorem 6.6.1, if one of the numbers u or v is positive and the other is negative, uv is negative and so is also not zero.

(3) The proof is essentially the same as the proof of Theorem 4.8.1. For example if $2a = 0$ then $a = 0$. Also $3(a - b) = 3a - 3b$.

Theorem 6.7.3. *Let u, v, and w be integers and $w \neq 0$. Then $u = v$ if and only if $uw = vw$.*

Proof. If $u = v$, then $uw = vw$, becaue $w = w$ and multiplication in I is not ambiguous by Theorem 6.3.1. Conversely, when $w \neq 0$ and $uw = vw$, $0 = uw - vw = (u - v)w$. But $w \neq 0$, so by Theorem 6.7.2, part (2), $u - v = 0$. Therefore $u = u - v + v = 0 + v$, and hence $u = v$.

For example, if $a = b$, then $5a = 5b$. Also if $3x = 3y$, then $x = y$.

It follows from this theorem that when $u \div v$ exists (and this can only happen when $v \neq 0$), it is unique; for if x and x' are integers such that $vx = u$ and $vx' = u$, then $vx = vx'$, and consequently by Theorem 6.7.3, since $v \neq 0$, $x = x'$.

We conclude this section with a theorem whose proof is left as an exercise.

Theorem 6.7.4. *Division on the right is distributive over addition and subtraction; that is, if u, v, and w are in I and $w \neq 0$, then*
$$(u + v) \div w = (u \div w) + (v \div w) \qquad \text{and} \qquad (u - v) \div w = (u \div w) - (v \div w).$$

For example, $(8 + 12) \div 4 = (8 \div 4) + (12 \div 4)$ and $(6 - 21) \div 3 = (6 \div 3) - (21 \div 3)$, as the reader can verify.

Exercises

1. Prove part (3) of Theorem 6.7.2 (see the proof of Theorem 4.8.1).
2. Prove that subtraction and division are not associative in I by proving the following:
   (a) There exist numbers u, v, and w in I such that $u - (v - w) \neq (u - v) - w$;
   (b) There exist numbers u, v, and w in I such that $u \div (v \div w) \neq (u \div v) \div w$.
3. (a) Use the distributive property—part (10) of Theorem 6.3.4—and then the definition of division to prove that if u, v, and w are integers and $w \neq 0$, then $[(u \div w) + (v \div w)]w = u + v$. Then notice from the definition of division in I that you have proved

   $$(u + v) \div w = (u \div w) + (v \div w).$$

   That is, you have proved part (1) of Theorem 6.7.4.
   (b) Similarly to (a), prove part (2) of Theorem 6.7.4. using part (3) of Theorem 6.7.2.
4. Prove that division is not distributive on the left over addition or subtraction, by showing that there exist numbers u, v, and w in I such that
   (a) $w \div (u + v) \neq (w \div u) + (w \div v)$;
   (b) $w \div (u - v) \neq (w \div u) - (w \div v)$.

6.8   Inequalities

Because the positive integers are the same as the positive whole numbers, the following theorem is just a restatement of Theorem 4.3.1, which we have already proved.

Theorem 6.8.1. *The set of positive integers is closed under addition and multiplication.*

We now have the necessary material to establish properties of inequalities for I. First of all, though, we must define $<$ (is less than) and $>$ (is greater than), which we do in the expected way.

Definition 6.8.1. *Let u and v be integers. Then $u < v$ and $v > u$ if and only if there exists a positive integer h such that $u + h = v$.*

Examples:

        (1) $8 < 10$, because $8 + 2 = 10$ and 2 is a positive integer.
        (2) $-3 < 2$, because $-3 + 5 = 2$ and 5 is a positive integer.
        (3) $-7 < -3$, because $-7 + 4 = -3$ and 4 is a positive integer.
        (4) $-2 < 0$, because $-2 + 2 = 0$ and 2 is a positive integer.
        (5) $-1 > -6$, $0 > -6$, and $3 > -6$.

Recalling that on the number line, when h is positive, $u + h$ is h units to the right of u, we see that $u < v$ if and only if u is to the left of v on the number line. Similarly, $u > v$ if and only if u is to the right of v on the number line.

Theorem 6.8.2. *If u and v are integers, then*

> (1) $u > v$ *if and only if $u - v$ is positive;*
> (2) $u < v$ *if and only if $u - v$ is negative.*

Proof.

(1) If $u > v$, then there exists a positive integer h such that $v + h = u$. But $h = u - v$ by definition of subtraction. Hence $u - v$ is positive. Conversely, if $u - v$ is positive, then, by definition of subtraction, $v + (u - v) = u$. Therefore, by definition of $>$, $u > v$.

(2) By part (1), $v > u$ if and only if $v - u$ is positive. $v > u$ is equivalent to $u < v$, and the statement that $v - u$ is positive is equivalent to the statement that $-(v - u) = u - v$ is negative. Therefore, $u < v$ if and only if $u - v$ is negative.

If we consider this theorem for the special case where $v = 0$, we get the following interesting information, which could also be obtained directly from the definition of $>$:

(1) $u > 0$ if and only if u is positive.

(2) $u < 0$ if and only if u is negative.

Theorem 6.8.2 also helps us get the trichotomy property for the integers, which we shall now state.

Theorem 6.8.3. *If u and v are integers, then one and only one of the following must hold:*

> (1) $u < v$;   (2) $u = v$;   (3) $u > v$.

Proof. Consider $u - v$, which by the closure property of I under subtraction [part (5) of Theorem 6.3.4] is an integer. Because $u - v$ is an integer one and only one of the following holds:

(a) $u - v$ is a negative integer,   (b) $u - v = 0$,   (c) $u - v$ is a positive integer.

Now (b) is equivalent to $u = v$ and by Theorem 6.8.2, (a) is equivalent to $u < v$ and (c) is equivalent to $u > v$. Consequently, one and only one of the following holds:

> (1) $u < v$,   (2) $u = v$.   (3) $u > v$,

and we have proved the theorem.

We now obtain several properties of $<$, some of which are similar to properties of the whole numbers obtained in Section 4.7.

Theorem 6.8.4. *Let u, v, and w be integers. Then*

(1) *If $u < v$ and $v < w$, then $u < w$;*
(2) *$u < v$ if and only if $u + w < v + w$;*
(3) *$u < v$ if and only if $u - w < v - w$;*
(4) *when $w > 0$, $u < v$ if and only if $wu < wv$;*
(5) *when $w < 0$, $u < v$ if and only if $wu > wv$.*

Proof.

(1) The proof of part (1) is essentially the same as the proof of Theorem 4.7.2 and is left as an exercise.

(2) If $u < v$ then by definition there exists a positive integer h such that $u + h = v$. Therefore, $u + h + w = v + w$; that is, $u + w + h = v + w$ where h is positive. Hence, by definition, $u + w < v + w$. Conversely, if $u + w < v + w$, there exists a positive integer h such that $u + w + h = v + w$. Therefore, $u + h + w = v + w$, and by Theorem 6.3.2 we can cancel the w to obtain $u + h = v$. Since h is positive, we have $u < v$.

(3) If $u < v$ then there exists a positive integer h such that $u + h = v$. Therefore $(u + h) - w = v - w$. Now, by Theorem 6.7.1, $(u + h) - w = (u - w) + h$, so $(u - w) + h = v - w$, and we have $u - w < v - w$. Conversely, if $u - w < v - w$, then, by part (2), $(u - w) + w < (v - w) + w$. But $(u - w) + w = u$ and $(v - w) + w = v$, and hence $u < v$.

(4) We shall use part (2) of Theorem 6.8.2, which says that an integer is less than a second integer if and only if the first minus the second is negative, to replace $u < v$ by "$u - v$ is negative" and $wu < wv$ by "$wu - wv$ is negative." That is, to prove that $u < v$ if and only if $wu < wv$, we shall prove that $u - v$ is negative if and only if $wu - wv = w(u - v)$ is negative. Since w is positive, and a product of two factors is negative if and only if one of the factors is positive and the other factor is negative, $u - v$ is negative if and only if $w(u - v)$ is negative. Therefore, when $w > 0$, $u < v$ if and only if $wu < wv$.

(5) Similarly to (4), we use both parts of Theorem 6.8.2 to replace $u < v$ by "$u - v$ is negative" and $wu > wv$ by "$wu - wv = w(u - v)$ is positive." Thus we shall prove that $u - v$ is negative if and only if $w(u - v)$ is positive. Since w is negative, and a product of two factors is positive if and only if both factors are negative or both factors are positive, it follows that $u - v$ is negative if and only if $w(u - v)$ is positive. Therefore, when $w < 0$, $u < v$ if and only if $wu > wv$.

It is important to notice that, by Theorem 6.8.4, when the same number is added to or subtracted from both sides of an inequality, the inequality remains unchanged in "direction." Also, when both sides are multiplied by the same positive number, the inequality remains in the same direction, but when both sides are multiplied by the same negative number, the direction of the inequality must be reversed.

Examples:

    (1) (a) $-5 < -2$ and $-2 < -1$, so $-5 < -1$.

       (b) If $x < -2$ and $-2 < y$, then $x < y$.

    (2) (a) $x - 3 < 2$ if and only if $x - 3 + 3 < 2 + 3$; that is, if and only if $x < 5$.

       (b) $x - 5 > -9$ if and only if $x - 5 + 5 > -9 + 5$; that is, if and only if $x > -4$.

       (c) $y + 3 < z + 3$ if and only if $y < z$.

    (3) (a) $x + 5 < 8$ if and only if $x + 5 - 5 < 8 - 5$; that is, if and only if $x < 3$.

       (b) $2x + 1 > 9$ if and only if $2x + 1 - 1 > 9 - 1$; that is, if and only if $2x > 8$, and this holds if and only if $x > 4$.

    (4) (a) $x < 4$ if and only if $2x < 8$, thus $2x < 8$ if and only if $x < 4$.

       (b) $x < -5$ if and only if $3x < -15$. Hence $3x < -15$ if and only if $x < -5$.

    (5) $x < -3$ if and only if $(-2)x > 6$ and hence $(-2)x > 6$ if and only if $x < -3$.

Exercises

    Explain why each of the statements 1–11 follows from Theorem 6.8.4. It is assumed that x is an integer.

1. $x - 4 < 6$ if and only if $x < 10$.   **2.** $x + 3 < 7$ if and only if $x < 4$.

3. $x - 2 > 5$ if and only if $x > 7$.   **4.** $x + 6 > -1$ if and only if $x > -7$.

5. $x < 3$ if and only if $5x < 15$.   **6.** $5x < 15$ if and only if $x < 3$.

7. $2x < -10$ if and only if $x < -5$.   **8.** $x < -3$ if and only if $-3x > 9$.

9. $-3x > 9$ if and only if $x < -3$.   **10.** $-x > 5$ if and only if $x < -5$.

11. $-4x > -12$ if and only if $x < 3$.

12. Find an integer x such that

    (a) $2x < x$;   (b) $x > 5x$;

    (c) $-x < -2$ but $3x \not< 6$;   (d) $-x > 5$ but $x \not> -5$;

    (e) $-2x < 4$ but $x \not< -2$;   (f) $-3x > 12$ but $x \not> -4$.

13. Complete each of the following by inserting $<$ or $>$ followed by a number:

    (a) $3x > 12$ if and only if x ___;   (b) $8x < -16$ if and only if x ___;

    (c) $-3x > 12$ if and only if x ___;   (d) $-3x < -12$ if and only if x ___;

    (e) $5x < 10$ if and only if $-x$ ___;   (f) $x > -3$ if and only if $2x$ ___;

    (g) $x < 6$ if and only if $4x$ ___;   (h) $-x > 2$ if and only if $7x$ ___.

chapter 7

Some elementary
number theory

7.1 Integers, primes, and composites

In this chapter we shall be concerned mainly with the set I of integers. It will be understood throughout this chapter that letters like a, b, c will denote integers. We shall extend to I the definition of the terms **divides**, **multiple**, and **divisor** given in Section 4.8.

Definition 7.1.1. *If a and b are integers, and there exists an integer h such that ah = b, we say that a **divides** b or that a is a **divisor** of b of that b is a* **multiple** *of a.*

We write $a|b$ for *a divides b* and $a \nmid b$ for *a does not divide b*.

Examples:

(1) $2|8$ because $2 \cdot 4 = 8$.
(2) $-2|8$ because $(-2) \cdot (-4) = 8$.
(3) $(-3)|(-6)$ because $(-3) \cdot 2 = -6$.
(4) $4|(-8)$ because $4 \cdot (-2) = -8$.
(5) $7|0$ because $7 \cdot 0 = 0$.
(6) $0|0$ because $0 \cdot 0 = 0$ or $0 \cdot 3 = 0$.
(7) $3 \nmid 5$ because there is no integer h such that $3h = 5$.

Notice that the divisors of 6 are 1, 2, 3, 6, -1, -2, -3, -6. These can be expressed as ± 1, ± 2, ± 3, ± 6 where $\pm$ means plus or minus and $+a$ means a. Some integers like 3 for example have only 4 divisors. The divisors of 3 are ± 1, ± 3. Although our formal definition will be worded slightly differently, these integers, except for ± 1, will be called **primes**.

Definition 7.1.2. *An integer $p \neq \pm 1$ that has no divisors different from ± 1 and $\pm p$ is called a* **prime***.*

For example 2, 3, 5, 7, 11, 13, 17 are primes, and so are -2, -3, -5, -7, -11, -13, -17. Another kind of integer is the **composite number.**

Definition 7.1.3. *An integer $n \neq 0$ such that $n = ab$ where a and b are integers, neither of which is ± 1, is called* **composite***.*

Examples:

    (1) 6 is composite because $6 = 3 \cdot 2$.
    (2) -15 is composite because $-15 = 5(-3)$.
    (3) 24 is composite because $24 = 6 \cdot 4$ or because $24 = 3 \cdot 8$.

Definition 7.1.4. *The divisors of 1, namely 1 and -1, are called* **units***.*

We have now classified the set I into four mutually disjoint subsets, namely

    1. $\{0\}$,
    2. $\{1, -1\}$, (the set of units),
    3. the set of primes,
    4. The set of composites.

To help understand some of the above ideas better, we will consider a different situation. Let us consider the set E of even integers. $E = \{\ldots, -6, -4, -2, 0, 2, 4, \ldots\} = \{x \in I \mid x = 2h$ for some $h \in I\} = \{2h \mid h \in I\}$. Restricting ourselves only to the set E, if a and b are in E, $a|b$ in E will mean that there exists a number h in E such that $ah = b$.

As far as being a prime relative only to the set E is concerned, we will consider a number p of E to be a *prime relative to E* if it has no divisors in E different from ± 1 and $\pm p$. We will consider a number n of E to be *composite relative to E* if $n = ab$ where a and b are in E and neither a nor b is ± 1. Of course, the integers ± 1 are not in E, so it is not really necessary to use ± 1 in these definitions of prime and composite relative to E. We do it only for the sake of generalizing the definitions in I.

Let us find some of the primes and composites relative to E. Notice that the element 6 of E is a prime relative to E, because 6 has no divisors in E different from ± 1 and ± 6. In fact, 6 has no divisors at all in E, because 6 cannot be expressed in the form $6 = ab$ where a and b are in E. In particular, $2 \nmid 6$ in E, because there is no number h in E such that $6 = 2h$.

Other examples:

(1) $8 = 4 \cdot 2$ in E, so 8 is composite and $2|8$ and $4|8$ in E.
(2) 2, 10, and 14 are primes in E, as are -2, -10, -14.
(3) $4 = 2 \cdot 2$ is composite relative to E.

Exercises

1. Name five primes and five composites of I not mentioned in this section.

2. Are the following prime or composite in I?
(a) 51;   (b) 91;   (c) 71;   (d) 123.

3. Find five primes and five composites in E relative to E (the set of even integers) not mentioned in this section.

4. List the divisors in E of each of the following numbers of E, and then decide which of the given numbers are primes and which are composite, relative to E:

$$14, 16, 18, 20, 22, 24, 26.$$

5. Let $T = \{x \in I \mid x = 3s + 1 \text{ where } s \in I\} = \{\ldots, -8, -5, -2, 1, 4, 7, 10, \ldots\}$.
(a) Define divisor, prime, composite, and unit in T (relative to T).
(b) Find five primes and five composites in T (relative to T).

6. Prove that there are infinitely many composite numbers in I.

7.2   The greatest common divisor

The integer 3 divides both 12 and 18 and is called a **common divisor**, of 12 and 18, according to the following definition.

Definition 7.2.1. *If $h|a$ and $h|b$, h is called a **common divisor** of a and b.*

That is, an integer that divides both a and b is called a common divisor of a and b.

The common divisors of 12 and 18 are ± 1, ± 2, ± 3, ± 6. The greatest of these is 6. Notice that every one of the common divisors of 12 and 18 is a divisor of 6, and hence 6 is the greatest common divisor of 12 and 18 according to the following definition.

Definition 7.2.2. *A whole number d is called **the greatest common divisor** of a and b if d is a common divisor of a and b and every common divisor of a and b divides d.*

We will use the symbol $g(a, b)$ to denote **the greatest common divisor** of a and b. Notice that $g(a, b)$ is a whole number, and hence must be zero or positive. Note also that we have not defined the greatest common divisor as the largest number which is a common divisor of a and b. The reason we have not is because our formalized definition will prove to be more convenient.

As another example, consider the numbers 24 and 36. The common divisors of these are ± 1, ± 2, ± 3, ± 4, ± 6, ± 12. In this case, $g(24, 36) = 12$, because 12 is a nonnegative common divisor of 12 and 36 such that every common divisor of 24 and 36 divides 12.

Can there be two different numbers, d and e, both of which are the greatest common divisor of a and b? No—because we can prove the next theorem.

Theorem 7.2.1. *The greatest common divisor of a and b is unique.*

Proof. Let d and e both be greatest common divisors of a and b; that is, let d and e both be nonnegative common divisors of a and b, such that every common divisor of a and b divides both d and e. We will show that $d = e$. Since e is a common divisor of a and b, $d|e$, so $e = d \cdot r$ for some integer r. Also, since d is a common divisor of a and b, $e|d$, so $d = e \cdot s$ for an integer s. Therefore,

$$d = es = drs.$$

Now if $d = 0$, then $e = d \cdot r = 0$, and $d = e$. If $d \neq 0$, then we may cancel the d to get $1 = r \cdot s$. Consequently $r = s = \pm 1$. Therefore $e = dr = d(\pm 1)$. Now since d and e are both positive, r cannot be -1, it must be 1. Therefore $d = e$.

This theorem justifies the word "the" in the name **the greatest common divisor.**

How do we find $g(a, b)$? A way of doing this is to find first all of the common divisors of a and b. This is not difficult if a and b are small, but it is difficult if they are large. One thing to realize is that an integer and its negative have the same divisors. For example, 4 and -4 have the same divisors, namely ± 1, ± 2, ± 4. Because of this, the greatest common divisor of two integers is equal to the greatest common divisor of the integers obtained when any minus signs occurring are removed. For example $g(-12, 18) = g(-12, -18) = g(12, -18) = g(12, 18)$.

If one of the numbers a or b is 0, the greatest common divisor is the other number, that is $g(r, 0) = r$. The proof of this follows from the definition of $g(a, b)$ and is left as an exercise. We have now reduced the consideration of finding $g(a, b)$ to the case where a and b are both positive. To reduce the problem still further, let us recall the division algorithm (Theorem 4.9.1). By the division algorithm, when a and b are both positive, there are unique whole numbers q and r such that $a = bq + r$ where $0 \le r < b$, and we will show that $g(a, b) = g(b, r)$.

Theorem 7.2.2. *If a and b are positive integers such that $a = bq + r$, then* $g(a, b) = g(b, r)$.

Proof. We will show that the common divisors of a and b are the same as the common divisors of b and r. To do this, let $h|a$ and $h|b$; thus $a = hs$ and $b = ht$ for integers t and s. Then

$$r = a - bq = hs - htq = h(s - tq),$$

so, by the definition of "divides," $h|r$, and h is a common divisor of b and r. Conversely, if $h|b$ and $h|r$, then $b = ht$ and $r = hw$ for integers t and w. Then

$$a = bq + r = htq + hw = h(tq + w),$$

so $h|a$, and h is a common divisor of a and b. Since a and b have the same common divisors as b and r, they have the same greatest common divisor.

This theorem helps because it enables us to reduce the problem of finding $g(a, b)$ to that of finding the greatest common divisor of smaller numbers. For example, take $g(221, 187)$. Now

$$221 = 187(1) + 34,$$

so $g(221, 187) = g(187, 34)$. But we can apply this technique again to get the numbers still smaller.

$$187 = 34(5) + 7,$$

and thus $g(187, 34) = g(34, 17)$. Applying the theorem again, we have

$$34 = 17(2) + 0,$$

hence $g(34, 17) = g(17, 0) = 17$. Therefore

$$g(221, 187) = g(187, 34) = g(34, 17) = g(17, 0) = 17.$$

This particular process of finding the greatest common divisor is called the **Euclidean Algorithm**. The last nonzero remainder is the greatest common divisor.

The Euclidean Algorithm can be used to find not only $g(a, b)$ but also integers s and t such that $g(a, b) = as + bt$.

For example, we have found that $g(221, 187) = 17$ from the equations

(1) $221 = 187(1) + 34,$

(2) $187 = 34(5) + 17,$

(3) $34 = 17(2) + 0.$

To find the integers s and t in the equation $17 = 221s + 187t$, we first express the remainder 34 of equation (1) as a multiple of 221 plus a multiple of 187. From (1), we obtain

$$(4) \qquad\qquad 34 = 221 - 187(1) = 221(1) + 187(-1).$$

We then use this to express the next remainder, which in this case is $g(221, 187)$, in this form. By (2),

$$17 = 187 - 34(5) = 187 + 34(-5).$$

But by (4),

$$34(-5) = [221(1) + 187(-1)](-5) = 221(-5) + 187(5).$$

Consequently

$$17 = 187 + 34(-5) = 187 + 221(-5) + 187(5) = 221(-5) + 187(6).$$

Hence in this case $s = -5$ and $t = 6$.

In general, we have the following theorem.

Theorem 7.2.3. *There exist integers s and t such that*

$$g(a, b) = as + bt.$$

Proof. The first remainder in the Euclidean Algorithm can be expressed as a times an integer plus b times an integer and then by means of this the second remainder can also be expressed in a similar form, etc.; finally, the last nonzero remainder (which is the greatest common divisor) can be expressed in a similar form.

Let us consider another example. Find $g(1050, 432)$, and find integers s and t such that $g(1050, 432) = 1050s + 432t$. The steps are as follows:

$$(1) \qquad\qquad 1050 = 432(2) + 186,$$
$$(2) \qquad\qquad 432 = 186(2) + 60,$$
$$(3) \qquad\qquad 186 = 60(3) + 6,$$
$$(4) \qquad\qquad 60 = 6(10) + 0.$$

Therefore $6 = g(1050, 432)$.

We will now express each of these remainders, starting with the first one, 186, and ending with the greatest common divisor, 6, in the form

$$1050 \text{ times an integer} + 432 \text{ times an integer}.$$

By equation (1),

(5) $$186 = 1050 - 432(2) = 1050(1) + 432(-2).$$

From (2) we obtain

(6) $$60 = 432 - 186(2) = 432 + 186(-2),$$

but by (5),

(7) $$186(-2) = [1050(1) + 432(-2)](-2) = 1050(-2) + 432(4).$$

Therefore, by (6) and (7),

(8) $$60 = 432 + 1050(-2) + 432(4) = 1050(-2) + 432(5).$$

From (3),

(9) $$6 = 186 - 60(3) = 186 + 60(-3).$$

But, from (8), we get

$$60(-3) = [1050(-2) + 432(5)](-3) = 1050(6) + 432(-15),$$

and combining this with (5), we have, from (9),

$$6 = 186 + 60(-3) = 1050(1) + 432(-2) + 1050(6) + 432(-15).$$

Consequently,

$$6 = 1050(7) + 432(-17).$$

Theorem 7.2.4. *Every integer greater than 1 is either a prime or can be expressed in at least one way as a product of positive primes.*

Proof. If $n > 1$ and n is not a prime, then $n = cd$ where both c and d are greater than 1 and less than n. Similarly, each of the integers c and d is either a prime or can be expressed as a product of two smaller positive integers, etc. We eventually arrive at an expression for n as a product of positive primes.

Notice that this theorem does not say that the expression of n as a product of primes is in any way unique. However, we shall prove uniqueness (except for the order of the factors) later. The theorem does imply, however, that every integer that is greater than 1 has a positive prime divisor, a fact that will be used in the next theorem.

There are infinitely many composite integers as one can easily see by noting that all of the even integers are composite. Are there infinitely many primes? Yes. there are, as we shall soon prove. Our proof is similar to one given by Euclid around 300 B.C.

Although Euclid was able to prove this over 2,000 years ago, no one has yet been able to prove that there are infinitely many "twin primes" or prove that there are only a finite number of them. Pairs of primes (such as 3 and 5, 5 and 7, 11 and 13, 17 and 19) which differ by 2 are called **twin primes**.

Another unsolved problem involving primes is **Goldbach's Conjecture** which can be stated as follows: Every even integer greater than 2 can be expressed as the sum of two primes. For example, $4 = 2 + 2$, $6 = 3 + 3$, $8 = 5 + 3$, $10 = 5 + 5 = 7 + 3$, $12 = 7 + 5$. No one has been able to prove that every even integer greater than 2 can be expressed as the sum of two primes or, the negation of this, that there is an even number greater than 2 that is not the sum of two primes.

Theorem 7.2.5. *There are infinitely many positive primes.*

Proof. To prove this, we will show that it is impossible to have only a finite number, say h, of positive primes. Assume there are only h positive primes, and let them be $p_1, p_2, \ldots, p_h$. Then consider their product plus 1. Let $n = p_1 p_2 \ldots p_h + 1$. This number n is greater than 1 and has no positive prime divisor because when n is divided by any of the primes the remainder is 1. But this cannot be, because, by Theorem 7.2.4, every integer greater than 1 has a positive prime divisor. Therefore, there cannot be only a finite number of positive primes and hence there are infinitely many.

Exercises

1. Prove directly from the definition of greatest common divisor that if a is any whole number, then $g(a, 0) = a$.

Use the Euclidean Algorithm to find the greatest common divisor of each of the following pairs of integers.

2. $a = 36$, $b = 76$;   **3.** $a = 299$, $b = 247$;   **4.** $a = 1001$, $b = 119$.

Express each of the following numbers as a product of positive primes.

5. 12;   **6.** 72;   **7.** 432;

8. 98;   **9.** 1010;   **10.** 91.

7.3   The fundamental theorem of arithmetic

Definition 7.3.1. *When $g(a, b) = 1$, a and b are said to be **relatively prime**.*

If a divides a product bc, does it have to divide either a or b? No, for example $6 \mid (4 \cdot 3)$ but $6 \nmid 4$ and $6 \nmid 3$. However, we can prove the following theorem.

Theorem 7.3.1. *If* $g(a, b) = 1$ *and* $a|bc$, *then* $a|c$.

Proof. By the Euclidean Algorithm, there exist integers s and t such that

$$1 = as + bt.$$

Multiply both sides of this equation by c, giving

$$c = asc + bct.$$

Now $a|bc$ so $bc = ah$ for some integer h. Hence:

$$c = asc + aht = a(sc + ht).$$

Therefore $a|c$, completing the proof.

Theorem 7.3.2. *If* p *is a prime and* $p|bc$, *then* $p|b$ *or* $p|c$.

Proof. If $g(p, b) = 1$ then, by Theorem 7.3.1, $p|c$. If $g(p, b) \neq 1$ then, since the only positive divisors of p are 1 and p, $g(p, b) = p$, so $p|b$.

It follows from this theorem that if $p|abc$, then $p|a$ or $p|bc$, and hence $p|a$ or $p|b$ or $p|c$. Similarly, if a prime divides a product of any number of factors, it must divide at least one of the factors. We will use this in the next theorem, which is called the **Fundamental Theorem of Arithmetic**. We will also use the fact that if p and q are primes and $p|q$, then $p = q$ because q has no positive divisors greater than 1, except q.

Theorem 7.3.3. (*The Fundamental Theorem of Arithmetic*) *Every integer that is greater than 1 is either a prime or can be expressed in a unique way* (*except for the order of the factors*) *as a product of positive primes.*

Proof. We have already proved in Theorem 7.2.4 that if n is an integer greater than 1 and is not a prime, it can be expressed in at least one way as a product of primes. We will now prove that such an expression is unique except for the order of the factors. To do this, let us assume that n can be expressed as a product of s positive primes, say $n = p_1 p_2 \ldots p_s$, and also as a product of t positive primes, say $n = q_1 q_2 \ldots q_t$ where $t \geq s$. We will prove that $s = t$ and that the p's are the same as the q's except possibly for the order in which they are written.

We have

(1) $$p_1 p_2 p_3 \cdots p_s = q_1 q_2 q_3 \cdots q_t.$$

Now $p_1 | q_1 q_2 \ldots q_t$ and so p_1 divides one of the q's; for convenience we may renumber the q's if necessary so that $p_1|q_1$. Since p_1 is a prime and q_1 is a prime, p_1 must equal q_1 and we can divide both sides of (1) by p_1, leaving one fewer primes on each side of the equation. We will then have

(2) $$p_2 p_3 \cdots p_s = q_2 q_3 \cdots q_t.$$

Now $p_2 | q_2 q_3 \ldots q_t$ and so p_2 divides one of the q's say q_2. As above, $p_2 = q_2$, and we divide both sides of (2) by p_2 to get

(3) $p_3 p_4 \cdots p_s = q_3 q_4 \cdots q_t$.

We then apply the same argument to p_3, p_4, etc., and since $t \geq s$ we can continue on through p_s. We will then have 1 on the left and, hence, cannot have any of the q's left over on the right. Thus $s = t$ and the p's and q's are the same except possibly for the order in which they are written. This completes the proof.

Examples:

$$(1) \qquad 50 = 2 \cdot 5 \cdot 5 = 2 \cdot 5^2.$$
$$(2) \qquad 360 = 2 \cdot 2 \cdot 2 \cdot 3 \cdot 3 \cdot 5 = 2^3 \cdot 3^2 \cdot 5.$$
$$(3) \quad 1,400 = 2 \cdot 2 \cdot 2 \cdot 5 \cdot 5 \cdot 7 = 2^3 \cdot 5^2 \cdot 7.$$
$$(4) \quad 98,000 = 2 \cdot 2 \cdot 5 \cdot 7 \cdot 2 \cdot 5 \cdot 7 \cdot 5 \cdot 2 = 2^4 \cdot 5^3 \cdot 7^2.$$

According to the theorem, the above factorizations are unique except for the order of factors.

It is interesting to notice that in the set E of even integers, factorization into primes is not unique. For example,

$$36 = 2 \cdot 18 = 6 \cdot 6.$$

In E, the numbers 2, 6, and 18 are all primes because they have no divisors in E, and thus we have two different factorizations of 36 into primes.

From the prime factorization of two integers, one can easily obtain their greatest common divisor. For example,

(1) $3500 = 2^2 \cdot 5^3 \cdot 7$ and $4400 = 2^4 \cdot 5^2 \cdot 11$.

We can get the prime factorization of $g(3500, 4400)$ as follows: As far as the prime 2 is concerned, $g(3500, 4400)$ must contain 2^2 as a factor; it could not contain 2^3 or 2^4 as a factor because then it would not divide $3500 = 2^2 \cdot 5^3 \cdot 7$. Now consider the prime 5. $g(3500, 4400)$ must contain 5^2 as a factor and cannot contain 5^3 as a factor for then it would not divide $4400 = 2^4 \cdot 5^2 \cdot 11$. It cannot contain any positive power of 7 as a factor for then it would not divide $4400 = 2^4 \cdot 5^2 \cdot 11$. It cannot contain any positive power of 11 as a factor for then it would not divide $3500 = 2^2 \cdot 5^3 \cdot 7$. Therefore, $g(3500, 4400) = 2^2 \cdot 5^2 = 100$.

In general, the greatest common divisor is obtained by comparing the exponents of each prime that occurs in both factorizations and taking the smaller of the two exponents for each such prime. Other examples:

(2) $12 = 2^2 \cdot 3$ and $30 = 2 \cdot 3 \cdot 5$, so $g(12, 30) = 2 \cdot 3 = 6$.
(3) $1,400 = 2^3 \cdot 5^2 \cdot 7$ and $98,000 = 2^4 \cdot 5^3 \cdot 7^2$, hence

$$g(1,400, 98,000) = 2^3 \cdot 5^2 \cdot 7.$$

Definition 7.3.2. *An integer h that is a multiple of each of the integers a_1, a_2, $\ldots a_k$ is called a **common multiple** of $a_1, a_2, \ldots, a_k$.*

For example 12 is a common multiple of 2, 6, and 3.

When one adds fractions, one usually does so by first obtaining the "lowest common denominator" of the fractions. For example, to add $\frac{5}{36} + \frac{7}{20} + \frac{11}{6}$, one first obtains the lowest common denominator 180. The number 180 is actually the **least common multiple** of 36, 20, and 6, according to the following definition.

Definition 7.3.3. *The whole number m that is a common multiple of a_1, a_2, ..., a_k, and which has the property that every common multiple of a_1, a_2, ..., a_k is a multiple of m, is called the **least common multiple** of $a_1, a_2, \ldots, a_k$.*

We shall use the notation $m(a_1, a_2, \ldots, a_k)$ to denote the least common multiple of $a_1, a_2, \ldots, a_k$. In particular, $m(a, b)$ denotes the least common multiple of a and b.

Some examples of the least common multiple are as follows:

(1) $36 = 2^2 \cdot 3^2$, $20 = 2^2 \cdot 5$, and $6 = 2 \cdot 3$. The least common multiple of 36, 20, 6 must contain $2^2 \cdot 3^2$ to be a multiple of 36, must contain $2^2 \cdot 5$ to be a multiple of 20, and must contain $2 \cdot 3$ to be a multiple of 6; so it must contain $2^2 \cdot 3^2 \cdot 5$. Now $2^2 \cdot 3^2 \cdot 5$ is a whole number and is a common multiple of 36, 20, 6. Also, every common multiple of $2^2 \cdot 3^2$, $2^2 \cdot 5$, $2 \cdot 3$ contains at least the factors 2^2, 3^2, and 5 and so is a multiple of $2^2 \cdot 3^2 \cdot 5$. Therefore, $m(36, 20, 6) = 2^2 \cdot 3^2 \cdot 5 = 180$.

(2) $1,400 = 2^3 \cdot 5^2 \cdot 7$ and $98,000 = 2^4 \cdot 5^3 \cdot 7^2$, hence $m(1,400, 98,000) = 2^4 \cdot 5^3 \cdot 7^2$.

(3) $3,500 = 2^2 \cdot 5^3 \cdot 7$, $605 = 5 \cdot 11^2$, and $200 = 2^3 \cdot 5^2$, and thus $m(3,500, 605, 200) = 2^3 \cdot 5^3 \cdot 7 \cdot 11^2 = 847,000$.

In general, to find the least common multiple of two or more integers, first express each integer by its prime factors and then take each prime as a factor, of the least common multiple, the greatest number of times that it appears in any one of the factorizations.

It is interesting to notice that it follows from this and the method of finding the greatest common divisor from the prime factorizations of integers that, for two integers a and b, $g(a, b) \cdot m(a, b) = ab$. For example, take $40 = 2^3 \cdot 5$ and $700 = 2^2 \cdot 5^2 \cdot 7$. Then $g(40, 700) = 2^2 \cdot 5$, $m(40, 700) = 2^3 \cdot 5^2 \cdot 7$, and $(2^2 \cdot 5)(2^3 \cdot 5^2 \cdot 7) = (2^3 \cdot 5)(2^2 \cdot 5^2 \cdot 7)$.

Thus, we can find $m(a, b)$ by first using the Euclidean Algorithm to find $g(a, b)$, and then using the equation $m(a, b) = \dfrac{ab}{g(a, b)}$. For example in Section 7.2 we used the Euclidean Algorithm to find $g(221, 187) = 17$. Hence,

$$m(221, 187) = \frac{221 \times 187}{17} = 2431.$$

This technique for finding $m(a, b)$ is especially useful when the prime factorizations of a and b are not easy to determine.

Exercises

Find the greatest common divisor and the least common multiple of the following pairs of numbers by using the prime factorizations of the given numbers. Leave your answers to 4 and 5 in factored form.

1. 72 and 81;   **2.** 336 and 72;

3. 72,000 and 18,000;   **4.** $2^2 \cdot 3^3 \cdot 5$ and $2^4 \cdot 3 \cdot 5^2 \cdot 7^3$;

5. $2^4 \cdot 3^5 \cdot 5^2 \cdot 7^3 \cdot 11$ and $2^5 \cdot 3^6 \cdot 5^4 \cdot 7 \cdot 11^2$.

Find the lowest common denominator for the following fractions.

6. $\dfrac{1}{14}, \dfrac{3}{20}, \dfrac{7}{72}$;

7. $\dfrac{1}{200}, \dfrac{3}{500}, \dfrac{1}{700}$;

8. $\dfrac{1}{32}, \dfrac{1}{24}, \dfrac{3}{14}, \dfrac{1}{20}$;

9. $\dfrac{1}{400}, \dfrac{1}{600}, \dfrac{1}{900}, \dfrac{1}{800}$.

10. Find $m(32, 24, 14, 20)$.

11. Find the least common multiple of each of the following pairs of numbers by first using the Euclidean Algorithm to find their greatest common devisor.

(a) 340, 119;   (b) 154, 56.

12. Find a number different from 36 in E that has two different factorizations into primes in E. Why are the numbers you used in your factorizations primes?

7.4   Congruence mod m

We shall now define a new kind of equality (equivalence relation) in the system I of integers. We shall then show that the reflexive, symmetric, and transitive properties hold and that addition and multiplication are not ambiguous with this equivalence relation. After that we will obtain some divisibility tests for integers and develop some tests for checking arithmetic calculations.

Definition 7.4.1. *If a, b, and m are integers, and $m > 1$, then a is congruent to b modulo m, and we write $a \equiv b$ mod m, if and only if $m \mid (a - b)$.*

Examples:

(1) $5 \equiv 2$ mod 3 because $3 \mid (5 - 2)$, that is $3 \mid 3$.

(2) $-1 \equiv 2$ mod 3 because $3 \mid (-1 - 2)$, that is $3 \mid -3$.

(3) $-8 \equiv 1$ mod 3 because $3 \mid (-8 - 1)$, that is $3 \mid -9$.

(4) $37 \equiv -3$ mod 10 because $10 \mid (37 - (-3))$, that is $10 \mid 40$.

(5) $-4 \equiv -40$ mod 9 because $9 \mid (-4 - (-40))$, that is $9 \mid 36$.

(6) $72 \equiv 0$ mod 9 because $9 \mid (72 - 0)$, that is $9 \mid 72$.

We will now show that for each particular $m > 1$, congruence mod m has the properties an "equality" needs to have, that is, that it is an **equivalence relation**.

Theorem 7.4.1. *If a, b, c, and m are integers and $m > 1$, then*
(1) $a \equiv a$ mod m *(Congruence* mod m *is reflexive);*
(2) *If $a \equiv b$ mod m, then $b \equiv a$ mod m (Congruence* mod m *is symmetric);*
(3) *If $a \equiv b$ mod m and $b \equiv c$ mod m, then $a \equiv c$ mod m (Congruence* mod m *is transitive).*

Proof.
(1) $a \equiv a$ mod m because $a - a = 0$ and $m \cdot 0 = 0$, so $m | 0$.
(2) If $a \equiv b$ mod m then $m | (a - b)$. Therefore $a - b = mh$ for some integer h. But then $b - a = -(a - b) = m(-h)$. Thus $m | (b - a)$ and therefore $b \equiv a$ mod m.
(3) If $a \equiv b$ mod m and $b \equiv c$ mod m, then $m | (a - b)$ and $m | (b - c)$. Hence, $a - b = mh$ and $b - c = mt$ for integers h and t. Therefore,

$$a - c = a - b + b - c = mh + mt = m(h + t).$$

Thus $m | (a - c)$, and we have $a \equiv c$ mod m.

For example,

(1) $8 \equiv 8$ mod 3;
(2) $11 \equiv 2$ mod 3 so $2 \equiv 11$ mod 3;
(3) $1110 \equiv 10$ mod 11 and $10 \equiv -1$ mod 11 therefore $1110 \equiv -1$ mod 11.

We will now see that addition, subtraction, and multiplication are not ambiguous mod m.

Theorem 7.4.2. *If $a \equiv a'$ mod m and $b \equiv b'$ mod m, then*
(1) $a + b \equiv a' + b'$ mod m;
(2) $a - b \equiv a' - b'$ mod m;
(3) $ab \equiv a'b'$ mod m.

Proof. If $a \equiv a'$ mod m and $b \equiv b'$ mod m, then $a - a' = mh$ and $b - b' = mt$ for integers h and t. Hence $a = a' + mh$ and $b = b' + mt$. Therefore

$$a + b = a' + b' + mh + mt,$$

so $(a + b) - (a' + b') = m(h + t)$ and we have $a + b \equiv a' + b'$ mod m. Also

$$a - b = a' + mh - b' - mt = a' - b' + m(h + t),$$

hence $(a - b) - (a' - b') = m(h + t)$ so $a - b \equiv a' - b' \mod m$. Similarly,

$$ab = (a' + mh)(b' + mt) = a'b' + mhb' + ma't + m^2 ht,$$

and therefore

$$ab - a'b' = m(hb' + a't + mht),$$

and we have $ab \equiv a'b' \mod m$, thus completing the proof.

For example, $10 \equiv 1 \mod 9$ and $-25 \equiv 2 \mod 9$, so

$$10 + (-25) \equiv 1 + 2 \equiv 3 \mod 9 \qquad \text{and} \qquad 10(-25) \equiv 1 \cdot 2 \equiv 2 \mod 9.$$

It follows from this theorem, that, in any calculation involving addition, multiplication, or subtraction, when numbers are replaced by numbers congruent to them mod m, the result of the calculation is congruent mod m to the result that would have been obtained if such replacements had not been made.

For example:

(1) $10 \equiv 1 \mod 3$, $5 \equiv -1 \mod 3$, and $9 \equiv 0 \mod 3$, so

$$10^2 \cdot 5 + 9^4 - 10(5) \equiv 1^2 \cdot (-1) + 0^4 - 1(-1) \equiv -1 + 0 + 1 \equiv 0 \mod 3.$$

(2) $10 \equiv 1 \mod 9$, so

$$\begin{aligned} 64{,}721 &= 6(10)^4 + 4(10)^3 + 7(10)^2 + 2(10) + 1 \\ &\equiv 6(1)^4 + 4(1)^3 + 7(1)^2 + 2(1) + 1 \mod 9 \\ &\equiv 6 + 4 + 7 + 2 + 1 \equiv 20 \equiv 2 \mod 9. \end{aligned}$$

(3) $10 \equiv -1 \mod 11$, so

$$\begin{aligned} 264{,}721 &= 2(10)^5 + 6(10)^4 + 4(10)^3 + 7(10)^2 + 2(10) + 1 \\ &\equiv 2(-1)^5 + 6(-1)^4 + 4(-1)^3 + 7(-1)^2 + 2(-1) + 1 \mod 11 \\ &\equiv -2 + 6 - 4 + 7 - 2 + 1 \equiv 6 \mod 11. \end{aligned}$$

This sort of sum where the signs alternate from $+$ to $-$ or from $-$ to $+$ is often called an "alternating sum."

(4) $10 \equiv 1 \mod 3$, so

$$4712 = 4(10)^3 + 7(10)^2 + 1(10) + 2 \equiv 4 + 7 + 1 + 2 \equiv 14 \equiv 5 \equiv 2 \mod 3$$

(5) $10 \equiv 0 \mod 5$, so $381 = 3(10)^2 + 8(10) + 1 \equiv 3(0)^2 + 8(0) + 1 \equiv 1 \mod 5$.

As in example (2), because $10 \equiv 1 \mod 9$, we can see that every number written in the decimal system is congruent mod 9 to the sum of its digits. Similarly, as in example (3), because $10 \equiv -1 \mod 11$ every number in the decimal system is congruent mod 11 to the "alternating sum" of its digits.

Also as in example (4), since $10 \equiv 1$ mod 3, every number written in the decimal system is congruent mod 3 to the sum of its digits.

Since the units digit always occurs with a plus sign when we get the "alternating sum" of the digits for a number in the decimal system of numerals by taking the number mod 11, it is usually easier to start writing the digits from the right. In Example (3), it is easier to write

$$264,721 \equiv 1 - 2 + 7 - 4 + 6 - 2 \equiv 6 \text{ mod } 11.$$

Examples:

    (1) $37254 \equiv 3 + 7 + 2 + 5 + 4 \equiv 21$ mod 9. Now $21 \equiv$ the sum of its own digits $\equiv 2 + 1 \equiv 3$ mod 9. Therefore $37254 \equiv 21 \equiv 3$ mod 9.

    (2) $137254 \equiv 4 - 5 + 2 - 7 + 3 - 1 \equiv -4 \equiv 7$ mod 11.

    (3) $6427 \equiv 7 - 2 + 4 - 6 \equiv 3$ mod 11.

    (4) $382 \equiv 3 + 8 + 2 \equiv 13 \equiv 4 \equiv 1$ mod 3.

Notice that $n \equiv 0$ mod m if and only if $m \mid (n - 0)$; that is, if and only if $m \mid n$. In particular, $n \equiv 0$ mod 9 if and only if $9 \mid n$; $n \equiv 0$ mod 3 if and only if $3 \mid n$; and $n \equiv 0$ mod 11 if and only if $11 \mid n$.

Using the information just obtained, we can say that a number in the base 10 numeral system is $\equiv 0$ mod 9 if and only if the sum of its digits is divisible by 9. Similar statements hold for 3 and 11. Thus we can now say:

    (1) *A number is divisible by 9 if and only if the sum of its digits is divisible by 9.*

    (2) *A number is divisible by 3 if and only if the sum of its digits is divisible by 3.*

    (3) *A number is divisible by 11 if and only if the "alternating sum" of its digits is divisible by 11.*

Also, because $10 \equiv 0$ mod 5, a number in the base 10 system is congruent to 0 modulo 5, if and only if its last digit is divisible by 5, and thus we can state:

    (4) *A number is divisible by 5 if and only if its units digit is 0 or 5.*

Examples:

    (1) Consider 31,823. The digit sum is 17 and the alternating digit sum is 11. Therefore it is not divisible by 3 or 9 but is divisible by 11. It is also not divisible by 5, because its last digit is neither 0 nor 5.

    (2) Consider 827,143,911. The digit sum is 36, and the alternating digit sum is 22. Therefore the given number is divisible by 3, 9, and 11. It is not divisible by 5.

As another example of an application of Theorem 7.4.2, let us consider the following:

$$3421 \equiv 3 + 4 + 2 + 1 \equiv 10 \equiv 1 \text{ mod } 9$$

$$2864 \equiv 2 + 8 + 6 + 4 \equiv 20 \equiv 2 \text{ mod } 9$$

Therefore, by Theorem 7.4.2, $3421 \times 2864 \equiv 1 \times 2 \equiv 2$ mod 9. But if one calculated 3421×2864 and got 9,897,744 for an answer, a mistake would be apparent if he checked and found it $\not\equiv 2$ mod 9. Now
$9,897,744 \equiv 9 + 8 + 9 + 7 + 7 + 4 + 4 \equiv 48 \equiv 4 + 8 \equiv 12 \equiv 3 \not\equiv 2$ mod 9.
Thus 9,897,744 is incorrect. If he repeats the problem and gets 9,797,744, then to check he obtains

$$9,797,744 \equiv 9 + 7 + 9 + 7 + 7 + 4 + 4 \equiv 47 \equiv 11 \equiv 2 \text{ mod } 9,$$

and, therefore, he has probably done the problem correctly, It is still possible that a mistake has been made, however, because if this answer differs from the correct one by a multiple of 9, it will still be $\equiv 2$ mod 9. The chances that an incorrect answer will differ from the correct one by a multiple of 9 are only 1 in 9, however. This checking process is called "casting out 9's."

One could also check by "casting out 11's." For example is $624 \times 231 = 144,214$? Now

$$624 \equiv 4 - 2 + 6 \equiv 8 \text{ mod } 11,$$
$$231 \equiv 1 - 3 + 2 \equiv 0 \text{ mod } 11.$$

Consequently,

$$624 \times 231 \equiv 8 \times 0 \equiv 0 \text{ mod } 11.$$

But $144,214 \equiv 4 - 1 + 2 - 4 + 4 - 1 \equiv 4 \not\equiv 0$ mod 11, so 144,214 is incorrect. Is 144,144 correct? $144,144 \equiv 4 - 4 + 1 - 4 + 4 - 1 \equiv 0$ mod 11 so it is probably correct. Since the correct answer is $\equiv 0$ mod 11, if 144,144 is incorrect, it differs from the correct answer by a multiple of 11. When an incorrect answer is obtained, however, there is only 1 chance in 11 that it differs from the correct answer by a multiple of 11.

Division problems can also be checked. For example, is $962 \div 37 = 26$? We check this by seeing whether $26 \times 37 = 962$, which can be checked by casting out 9's or 11's or both. If it checks with both, it is probably correct, and in fact it is either correct or differs from the correct answer by a multiple of 99. However, there is only 1 chance in 99 that an incorrect answer differs from the correct answer by a multiple of 99. If it does not check with one method but does with the other, then it is incorrect.

Addition problems can be checked too. For example, is the following correct?

342	$342 \equiv 3 + 4 + 2 \equiv 9 \equiv 0$ mod 9.
685	$685 \equiv 6 + 8 + 5 \equiv 19 \equiv 10 \equiv 1$ mod 9.
907	$907 \equiv 9 + 0 + 7 \equiv 16 \equiv 7$ mod 9.
1924	The correct sum $\equiv 0 + 1 + 7 \equiv 8$ mod 9.

Now $1924 \equiv 1 + 9 + 2 + 4 \equiv 16 \equiv 7 \not\equiv 8$ mod 9, and hence a mistake has been made in the addition.

In the base 5 numeral system, $5 \equiv 1 \bmod 4$ and $5 \equiv -1 \bmod 6$, so

$$(3412)_5 = 3(5)^3 + 4(5)^2 + 1(5) + 2 \equiv 3(1)^3 + 4(1)^2 + 1(1) + 2$$
$$\equiv 3 + 4 + 1 + 2 = 10 = (20)_5 \equiv 2 \bmod 4,$$

and

$$(3412)_5 \equiv 3(-1)^3 + 4(-1)^2 + 1(-1) + 2 \equiv -3 + 4 - 1 + 2 \equiv 2 \bmod 6.$$

One would check arithmetic in the base 5 numeral system by casting out 4's or 6's, because a number in this system is congruent mod 4 to the sum of its digits and is congruent mod 6 to the alternating sum of its digits. In the base 3 numeral system one would check by casting out 2's or 4's. For example, suppose we are trying to multiply $(234)_5 \times (123)_5$ and obtain $(44442)_5$ as the answer. Let us check this result by first casting out 4's and then by casting out 6's.

$$(234)_5 \equiv 2 + 3 + 4 \equiv (14)_5 \equiv 1 + 4 \equiv (10)_5 \equiv 1 + 0 \equiv 1 \bmod 4.$$
$$(123)_5 \equiv 1 + 2 + 3 \equiv (11)_5 \equiv 1 + 1 \equiv (2)_5 \equiv 2 \bmod 4.$$

Therefore if $(44442)_5$ is the correct result, we must have $(44442)_5 \equiv 1 \times 2 \equiv 2$ mod 4. Now

$$(44442)_5 \equiv 4 + 4 + 4 + 4 + 2 \equiv (33)_5 \equiv 3 + 3 \equiv (11)_5 \equiv 1 + 1 \equiv 2 \bmod 4.$$

Since $(44442)_5 \equiv 2 \bmod 4$, it checks, and is either correct or differs from the correct answer by a multiple of 4.

Now let us check by casting out 6's.

$$(234)_5 \equiv 4 - 3 + 2 \equiv 3 \bmod 6.$$
$$(123)_5 \equiv 3 - 2 + 1 \equiv 2 \bmod 6.$$

Therefore, if our result is correct, $(44442)_5 \equiv 3 \times 2 \equiv (11)_5 \equiv 1 - 1 \equiv 0$ mod 6. But

$$(44442)_5 \equiv 2 - 4 + 4 - 4 + 4 \equiv 2 \not\equiv 0 \bmod 6,$$

and therefore $(44442)_5$ is definitely the wrong answer.

In ordinary "everyday" arithmetic in the decimal system, one usually checks only by casting out 9's. It is a good, quick checking technique. It is important to remember, however, that when the procedure is correctly performed and

    a) *an answer checks, it is quite likely to be correct and is either correct or differs from the correct answer by a multiple of 9;*
    b) *an answer does not check, it is definitely incorrect.*

Exercises

Test each of the following for divisibility by (a) 3; (b) 5; (c) 9; (d) 11.

1. 40830;   **2.** 20117936;   **3.** 350724;

4. 55166155011;   **5.** 812368.

Perform the indicated operations and check by casting out 9's.

6. 324 × 846;   **7.** 603 × 179;

8. 3482 × 83;   **9.** 364 ÷ 26;

10. 8320 + 729 + 964 + 2379;   **11.** 23,718 − 1,846.

Check the following by casting out 9's and 11's. If your check indicates an error, do the calculation correctly and then check again.

12. 712 × 375 = 267,000;   **13.** 423 × 86 = 37,278;

14. 482 + 1,796 + 347 = 2,725;

15. 3,008 + 47,982 + 61,706 + 8,241 = 131,837;

16. 901 ÷ 17 = 53;   **17.** 14,283 ÷ 23 = 632.

Check the following by casting out 4's and 6's. If your check indicates an error, perform the operations correctly, and then recheck.

18. $(14143)_5 \div (32)_5 = (232)_5$;

19. $(321)_5 \times (432)_5 = (32012)_5$;

20. $(241)_5 + (324)_5 + (120)_5 = (1243)_5$.

chapter 8

The rational numbers

8.1 Introduction to rational numbers

When we defined division in I, we noticed that I is not closed under division, because division by zero has no meaning. We also noticed that even with division by zero excluded, I is not closed under division. For example, $11 \div 2$ does not exist in I, because there is no integer x such that $2x = 11$. In order to remedy this situation, we shall develop the system of **rational numbers**, also called the system of **fractions**. Then there will be numbers (not necessarily in I) for $a \div b$ for all integers a and b as long as $b \neq 0$.

Another situation indicating the need for fractions occurs in measuring. When one measures, say, a pencil, he is likely to do so by placing it along the edge of a ruler. This amounts, essentially, to placing it on the number line with one end at 0 and seeing where the other end is. If the other end is exactly at 5, the pencil is 5 units long. However, the end may not be at 5, or any other point representing an integer. In Figure 8.1.1, the point of the pencil is about midway between 5 and 6. The number representing its length in the units of the given number line is thus not an integer. It is clear that the integers are important but inadequate for measuring. We need to have "fractions."

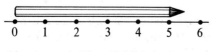

Figure 8.1.1

127

Of course, all of us are familiar with fractions and know that the pencil in the diagram is about $5\frac{1}{2}$ or $\frac{11}{2}$ units long and that the number $\frac{11}{2}$ or $5\frac{1}{2}$ is $11 \div 2$, but do we know what $\frac{11}{2}$ means? Do we know why $\frac{11}{2}$, $5\frac{1}{2}$, and $11 \div 2$ are equivalent? Discussion of these ideas will be included in this chapter, in our continuation of the development of the number system.

When we introduced the integers, the motivation was to enlarge the system of whole numbers to a system in which subtraction was always possible. The reason for defining equality, addition, and multiplication in I as we did was to make (a, b) turn out to be $a - b$.

In order to enlarge the system of integers to a system in which division (except by zero) is always possible, we consider the Cartesian product $F = I \times N$ of the set I of integers with the set N of nonzero integers. To help distinguish the elements of F from the ordered pairs of $I = W \times W$ we shall use square brackets rather than parentheses for the ordered pairs of F. Thus,

$$F = I \times N = \{[a, b] \mid a \in I \text{ and } b \in N\}.$$

We shall want $[a, b]$ to turn out to be $a \div b$ and our definitions of equality, addition, and multiplication in F will be so motivated.

Definition 8.1.1. (*Definition of equality in F*). *Two elements* $[a, b]$ *and* $[c, d]$ *of F are called equal, and we write* $[a, b] = [c, d]$, *if and only if* $ad = bc$ *in I.*

For example, $[11, 2]$, $[2, -3]$, $[0, 2]$, $[-4, 6]$ are in F, and $[2, -3] = [-4, 6]$, because $2 \cdot 6 = (-3)(-4)$. Similarly $[0, 2] = [0, -3]$, because $0 \cdot (-3) = 2 \cdot 0$ in I.

We now define addition and multiplication in F.

Definition 8.1.2. (*Definition of addition and multiplication in F.*) *Let* $[a, b]$ *and* $[c, d]$ *be any elements of F. Then*

$$[a, b] + [c, d] = [ad + bc, bd], \quad and \quad [a, b][c, d] = [ac, bd].$$

For example, $[2, 3] + [5, 4] = [2 \cdot 4 + 3 \cdot 5, 3 \cdot 4] = [23, 12]$ and $[2, 3][-5, 4] = [2(-5), 3 \cdot 4] = [-10, 12]$.

Definition 8.1.3. *The set* $F = I \times N$ *with the equality, addition, and multiplication of Definitions 8.1.1 and 8.1.2 is called the* **system of rational numbers**.

For any rational number $[a, b]$, the first component a is called the numerator and the second component b is called the denominator. For convenience of notation, a rational number $[a, b]$ is usually written as $\dfrac{a}{b}$ or as a/b. It should be remembered, however, that a/b is an ordered pair of integers with $b \neq 0$ and that $a/b = c/d$ means $ad = bc$. In particular, we will write $\frac{11}{2}$ for the rational number $[11, 2]$ and realize that $\frac{21}{35} = \frac{12}{20}$, because $21 \cdot 20 = 35 \cdot 12$. It is worth emphasizing that if a/b is a rational number, then this implies that a and b are integers and $b \neq 0$.

In order to have the necessary freedom when using equalities of rational numbers, we prove the following theorem indicating that equality of fractions is an **equivalence relation**.

Theorem 8.1.1. *If* $\dfrac{a}{b}, \dfrac{c}{d}$, *and* $\dfrac{g}{h}$ *are rational numbers, then*

(1) $\dfrac{a}{b} = \dfrac{a}{b}$ *(Equality of rational numbers is reflexive);*

(2) *If* $\dfrac{a}{b} = \dfrac{c}{d}$, *then* $\dfrac{c}{d} = \dfrac{a}{b}$ *(Equality of rational numbers is symmetric);*

(3) *If* $\dfrac{a}{b} = \dfrac{c}{d}$ *and* $\dfrac{c}{d} = \dfrac{g}{h}$, *then* $\dfrac{a}{b} = \dfrac{g}{h}$ *(Equality of rational numbers is transitive).*

Proof of (1). $a/b = a/b$ because $ab = ba$.

Proof of (2). If $a/b = c/d$ then $ad = bc$. Therefore, because multiplication is commutative in I and because equality is symmetric in I, $cb = da$ and, hence, $c/d = a/b$.

Proof of (3). If $a/b = c/d$ and $c/d = g/h$, then by definition of equality $ad = bc$ and $ch = dg$. Therefore, $adh = bch$ and $bch = bdg$, so $adh = bdg$ because equality is transitive in I. Now $d \neq 0$, since it is a denominator, and we can use the cancellation property (Theorem 6.7.3) to cancel the d to get $ah = bg$. Hence, $a/b = g/h$ by definition of equality in F.

We now have the following theorems, proofs of which are left as exercises.

Theorem 8.1.2. *If* $\dfrac{a}{b}$ *is a rational number and* k *is a nonzero integer, then*

$$\frac{a}{b} = \frac{ka}{kb}.$$

For example: $\dfrac{2}{3} = \dfrac{2 \cdot 2}{2 \cdot 3} = \dfrac{4}{6}$; $\qquad \dfrac{15}{20} = \dfrac{5 \cdot 3}{5 \cdot 4} = \dfrac{3}{4}$;

$\qquad\qquad \dfrac{21}{27} = \dfrac{7}{9}$; $\qquad \dfrac{-3}{-4} = \dfrac{(-1)(3)}{(-1)(4)} = \dfrac{3}{4}$.

Theorem 8.1.3. *If* c *and* d *are nonzero integers, then* $\dfrac{0}{c} = \dfrac{0}{d}$.

For example, $\dfrac{0}{3} = \dfrac{0}{7} = \dfrac{0}{-6} = \dfrac{0}{-1} = \dfrac{0}{1}$.

Whenever the numerator and denominator have a greatest common divisor d not equal to 1, the d can be cancelled so that the fraction is equal to one whose numerator and denominator are relatively prime. If the numerator and denominator are relatively prime, the fraction is said to be in the **lowest terms**. Theorem 8.1.2 is often used to reduce a fraction to lowest terms. For example

$$\frac{18}{27} = \frac{9 \cdot 2}{9 \cdot 3} = \frac{2}{3}, \quad \text{and} \quad \frac{-6}{8} = \frac{2(-3)}{2 \cdot 4} = \frac{-3}{4}.$$

Exercises

1. Use the definition of equality in F to prove that

$$\frac{3}{2} \neq \frac{5}{3}, \quad \frac{0}{3} \neq \frac{3}{b}, \quad \frac{-3}{2} = \frac{3}{-2}, \quad \text{and} \quad \frac{-1}{-2} = \frac{1}{2}.$$

2. Use the definition of equality in F to prove Theorem 8.1.2.

3. Use the definition of equality in F to prove Theorem 8.1.3.

4. Reduce to lowest terms: (a) $\dfrac{36}{32}$; (b) $\dfrac{126}{360}$; (c) $\dfrac{27}{-8}$; (d) $\dfrac{242}{-96}$.

8.2   Addition and multiplication of rational numbers

Now that we have defined rational numbers and have a notation for them, let us consider adding and multiplying them. According to Definition 8.1.2 and our notation for rational numbers, we have

$$\frac{a}{b} + \frac{c}{d} = \frac{ad + bc}{bd} \quad \text{and} \quad \frac{a}{b} \times \frac{c}{d} = \frac{ac}{bd}$$

for any two rational numbers $\dfrac{a}{b}$ and $\dfrac{c}{d}$.

For example,

(1) $\dfrac{2}{3} + \dfrac{3}{4} = \dfrac{2 \cdot 4 + 3 \cdot 3}{3 \cdot 4} = \dfrac{8 + 9}{12} = \dfrac{17}{12}.$

(2) $\dfrac{-1}{2} + \dfrac{3}{-4} = \dfrac{(-1)(-4) + 2 \cdot 3}{2(-4)} = \dfrac{4 + 6}{-8} = \dfrac{10}{-8} = \dfrac{5}{-4}.$

(3) $\dfrac{2}{3} \times \dfrac{3}{4} = \dfrac{2 \cdot 3}{3 \cdot 4} = \dfrac{1 \cdot 2 \cdot 3}{3 \cdot 2 \cdot 2} = \dfrac{1}{2}.$

(4) $\dfrac{0}{5} \times \dfrac{-3}{6} = \dfrac{0(-3)}{5 \cdot 6} = \dfrac{0}{30} = \dfrac{0}{1}.$

(5) $\dfrac{1}{2} + \dfrac{1}{4} = \dfrac{1 \cdot 4 + 2 \cdot 1}{2 \cdot 4} = \dfrac{6}{2 \cdot 4} = \dfrac{3}{4}.$

It should be noted that we do not prove that rational numbers add and multiply this way. They add and multiply this way by definition.

In Sections 4.2 and 6.3, we saw that addition and multiplication of whole numbers and integers are **well defined**; that is, they are not ambiguous. We shall now see that this is also the case in F. It is necessary to be concerned with this, because there are so many ways of expressing each rational number. Before giving a proof for fractional numbers, however, we will first consider a numerical example.

According to the definition of equality of rational numbers.

$$\frac{2}{6} = \frac{3}{9} \quad \text{and} \quad \frac{5}{10} = \frac{4}{8},$$

because $2 \cdot 9 = 6 \cdot 3$ and $5 \cdot 8 = 10 \cdot 4$. Now if we add $\frac{2}{6} + \frac{5}{10}$ will we get a number equal to that obtained when we add $\frac{3}{9} + \frac{4}{8}$? Let us see.

$$\frac{2}{6} + \frac{5}{10} = \frac{2 \cdot 10 + 6 \cdot 5}{6 \cdot 10} = \frac{20 + 30}{60} = \frac{50}{60},$$

$$\frac{3}{9} + \frac{4}{8} = \frac{3 \cdot 8 + 9 \cdot 4}{9 \cdot 8} = \frac{24 + 36}{72} = \frac{60}{72}.$$

We do get equal results because $\frac{50}{60}$ does equal $\frac{60}{72}$ since $50 \times 72 = 60 \times 60 = 3600$, so at least for this example addition is not ambiguous. What about multiplication for this example?

$$\frac{2}{6} \times \frac{5}{10} = \frac{10}{60} \quad \text{and} \quad \frac{3}{9} \times \frac{4}{8} = \frac{12}{72}.$$

Again the results are equal, because $\frac{10}{60} = \frac{12}{72}$ since $10 \times 72 = 60 \times 12 = 720$.

In general we have the following theorem, which shows that addition and multiplication of rational numbers is **well defined** and hence is not ambiguous.

Theorem 8.2.1. *If* $\dfrac{a}{b} = \dfrac{a'}{b'}$ *and* $\dfrac{c}{d} = \dfrac{c'}{d'}$ *are pairs of equal rational numbers, then*

(1) $\dfrac{ad + bc}{bd} = \dfrac{a'd' + b'c'}{b'd'}$

(2) $\dfrac{ac}{bd} = \dfrac{a'c'}{b'd'}.$

Proof. According to the definition of equality in F, we must show that $(ad + bc)b' d' = bd(a' d' + b' c')$ and $acb' d' = bda' c'$ both hold if $ab' = ba'$ and $cd' = dc'$. Now, by using the distributive and commutative properties and then replacing ab' by ba' and cd' by dc', we obtain

$$(ad + bc)b' d' = adb' d' + bcb' d' = (ab')(dd') + (cd'))bb')$$
$$= (ba')(dd') + (dc')(bb').$$

Similarly

$$bd(a' d' + b'c') = bda' d' + bdb' c' = (ba')(dd') + (dc')(bb').$$

Therefore $(ad + bc)b' d' = bd(a' d' + b' c')$, and we have

$$\frac{ad + bc}{bd} = \frac{a' d' + b' c'}{b' d'}.$$

To prove the second part of Theorem 8.2.1, we again use the commutative property, with $ab' = ba'$ and $cd' = dc'$, to obtain $acb' d' = (ab')(cd') = (a' b)(c' d) = bda' c'$. Consequently

$$\frac{ac}{bd} = \frac{a' c'}{b' d'}.$$

The familiar way of adding fractions by getting a " common denominator " is illustrated as follows:

Example 1. $\dfrac{1}{2} + \dfrac{1}{4} = \dfrac{2}{4} + \dfrac{1}{4} = \dfrac{3}{4}.$

Example 2. $\dfrac{5}{6} + \dfrac{3}{8} = \dfrac{40}{48} + \dfrac{18}{48} = \dfrac{58}{48} = \dfrac{29}{24}.$

Example 3. $\dfrac{5}{6} + \dfrac{3}{8} = \dfrac{20}{24} + \dfrac{9}{24} = \dfrac{29}{24}.$

Example 4. $\dfrac{-7}{15} + \dfrac{1}{6} = \dfrac{-14}{30} + \dfrac{5}{30} = \dfrac{-9}{30} = \dfrac{-3}{10}.$

In Examples 3 and 4, the "lowest common denominator" was used. The "lowest common denominator" is actually the least common multiple, of the denominators as we saw in Section 7.3. It is usually the simplest denominator to use when adding rational numbers.

The following theorem shows why we can simply " add the numerators " when two rational numbers have the same denominator.

Theorem 8.2.2. *If $\dfrac{a}{c}$ and $\dfrac{b}{c}$ are rational numbers with the same denominator,*
then

$$\frac{a}{c} + \frac{b}{c} = \frac{a+b}{c}.$$

Proof. By Definition 8.1.2 for addition in F and by Theorem 8.1.2,

$$\frac{a}{c} + \frac{b}{c} = \frac{ac + bc}{cc} = \frac{c(a+b)}{cc} = \frac{a+b}{c}.$$

We are accustomed to replacing $\frac{3}{1}$ by 3, $\frac{0}{1}$ by 0, $-2/1$ by -2, and in general, $n/1$ by n for every integer n. In the next theorem, we shall see why this can be done. Actually, according to our definitions $n/1$ and n are not conceptually the same, but we shall prove that the set $I' = \{n/1 \mid n \in I\}$ of all rational numbers with denominator 1 is isomorphic under addition anp multiplication to I. Thus we shall see that I and I' behave the same under equality, addition, and multiplication.

Theorem 8.2.3. *The* 1–1 *correspondence* $n \leftrightarrow n' = \dfrac{n}{1}$ *of* I *and* $I' = \left\{\dfrac{n}{1} \,\middle|\, n \in I\right\}$
is an isomorphism under addition and multiplication.

Proof. The pairing $n \leftrightarrow n' = n/1$ is clearly a 1–1 correspondence of I and $I' = \{n/1 \mid n \in I\}$, and we shall show that (1) $a = b$ if and only if $a' = b'$, (2) $(a+b)' = a' + b'$, and (3) $(ab)' = a'b'$.

(1) By definition of equality in F, $a = b$ if and only if $a/1 = b/1$. Therefore $a = b$ if and only if $a' = b'$.

(2) According to our notation for the $1-1$ correspondence,

$$(a+b)' = \frac{(a+b)}{1} \qquad \text{and} \qquad a' + b' = \frac{a}{1} + \frac{b}{1}.$$

But

$$\frac{a}{1} + \frac{b}{1} = \frac{a+b}{1}$$

by Theorem 8.2.2. and hence $(a+b)' = a' + b'$.

(3) $(ab)' = ab/1$ and $a'b' = a/1 \cdot b/1$, but by definition of multiplication in F, $a/1 \cdot b/1 = ab/1$ and, therefore $(ab)' = a'b'$.

Because I and I' behave the same under equality, addition, and multiplication we shall identify I' with I and usually write a for $a/1$ for all a in I.

We shall also write $a = a/1$. Thus we will consider that I is a subset of F. We then have $W \subset I \subset F$.

Exercises

For exercises 1–4, find the lowest common denominator for the given fractions:

1. $\dfrac{1}{14}, \dfrac{-3}{20}, \dfrac{7}{72}$;

2. $\dfrac{1}{200}, \dfrac{-3}{500}, \dfrac{1}{-700}$;

3. $\dfrac{1}{32}, \dfrac{-1}{24}, \dfrac{3}{-14}$;

4. $\dfrac{-3}{400}, \dfrac{1}{-600}, \dfrac{1}{-900}, \dfrac{1}{800}$.

Find the following sums and products, expressing your results in lowest terms.

5. $\dfrac{1}{2} + \dfrac{1}{3}$;

6. $\left(\dfrac{1}{6} + \dfrac{1}{8}\right) + \dfrac{1}{10}$;

7. $\dfrac{3}{8} + \left(\dfrac{7}{12} + \dfrac{1}{10}\right)$;

8. $\left(\dfrac{3}{8} + \dfrac{7}{12}\right) + \dfrac{1}{10}$;

9. $\dfrac{3}{4} + \left(\dfrac{5}{8} + \dfrac{9}{16}\right)$;

10. $\dfrac{2}{3} \times \left(\dfrac{3}{4} + \dfrac{7}{10}\right)$;

11. $\left(\dfrac{3}{4} + \dfrac{5}{8}\right) + \dfrac{9}{16}$;

12. $\left(\dfrac{21}{24} + \dfrac{7}{36}\right) \times \dfrac{3}{8}$;

13. $\dfrac{21}{32} \times \dfrac{16}{15}$;

14. $\dfrac{26}{51} + \dfrac{15}{19}$;

15. $\dfrac{-3}{20} + \left(\dfrac{1}{14} + \dfrac{7}{2}\right)$;

16. $\left(\dfrac{-5}{14} + \dfrac{7}{4}\right) + \dfrac{-3}{20}$.

Let S be the set of all positive integers. Determine whether the correspondence $n \leftrightarrow n'$ given for each S' is an isomorphism under (a) addition, (b) multiplication.

17. $n \leftrightarrow n' = \dfrac{1}{n}$ for all n in S, where $S' = \left\{\dfrac{1}{n} \,\middle|\, n \in S\right\}$;

18. $n \leftrightarrow n' = \dfrac{3}{n}$ for all n in S, where $S' = \left\{\dfrac{3}{n} \,\middle|\, n \in S\right\}$.

8.3   Subtraction and division of rational numbers

Our definitions of subtraction and division for rational numbers will be essentially the same as they are for whole numbers.

Definition 8.3.1. *If f and g are rational numbers, then $f - g$ is the number x such that $g + x = f$.*

Definition 8.3.2. *If f and g are rational numbers, then $f \div g$ is the number x such that $gx = f$.*

Notice that by Definition 8.3.1, $g + (f - g) = f$ and $(f + h) - h = f$, and that by Definition 8.3.2, $g(f \div g) = f$ and $(fh) \div h = f$. These facts are important and will be used later.

These definitions do not explain how one actually subtracts or divides rational numbers, so let us investigate further.

We shall first consider subtraction. Take for example $\frac{5}{7} - \frac{2}{3}$. This is the number x (if there is one) such that $\frac{2}{3} + x = \frac{5}{7}$. To determine if there is such a number, it is convenient to use a common denominator (in this case 21) and then try to find an appropriate x of the form $y/21$ where y is an integer. Now $\frac{2}{3} = \frac{14}{21}$ and $\frac{5}{7} = \frac{15}{21}$ so we are trying to find y such that

$$\frac{14}{21} + \frac{y}{21} = \frac{15}{21}.$$

We can see that $\frac{1}{21}$ is the correct number. That is

$$\frac{2}{3} + \frac{1}{21} = \frac{14}{21} + \frac{1}{21} = \frac{15}{21} = \frac{5}{7}.$$

Hence, by the definition of subtraction, $\frac{5}{7} - \frac{2}{3} = \frac{1}{21}$.

To see how to make such subtractions in every case, we consider $a/b - c/d$. This is the number x (if there is one) such that $c/d + x = a/b$. To determine if there is such a number, it is convenient to use a common denominator (in this case bd) and then try to find an appropriate x of the form y/bd where y is an integer. Now $a/b = ad/bd$ and $c/d = bc/bd$, so we are trying to find y such that

$$\frac{bc}{bd} + \frac{y}{bd} = \frac{ad}{bd}.$$

From the definition of subtraction in I, we see that $ad - bc$ is the correct number. That is

$$\frac{c}{d} + \frac{ad - bc}{bd} = \frac{bc}{bd} + \frac{ad - bc}{bd} = \frac{bc + (ad - bc)}{bd} = \frac{ad}{bd} = \frac{a}{b}.$$

Hence by the definition of subtraction,

$$\frac{a}{b} - \frac{c}{d} = \frac{ad - bc}{bd}.$$

and we have proved the following theorem:

Theorem 8.3.1. *If $\dfrac{a}{b}$ and $\dfrac{c}{b}$ are rational numbers, then*

$$\frac{a}{b} - \frac{c}{d} = \frac{ad - bc}{bd}.$$

Notice how similar this looks to addition. For example

$$\frac{3}{4} - \frac{2}{5} = \frac{3 \cdot 5 - 4 \cdot 2}{4 \cdot 5} = \frac{15 - 8}{20} = \frac{7}{20}$$

As in the case of addition, it is often easiest to get a common denominator first and then subtract.

Theorem 8.3.2. *If $\dfrac{a}{c}$ and $\dfrac{b}{c}$ are fractions with the same denominator, then*

$$\frac{a}{c} - \frac{b}{c} = \frac{a - b}{c}.$$

Proof. It follows from the previous theorem and Theorem 8.1.2 that

$$\frac{a}{c} - \frac{b}{c} = \frac{ac - cb}{cc} = \frac{c(a - b)}{cc} = \frac{a - b}{c}.$$

For example, $\dfrac{3}{8} - \dfrac{1}{6} = \dfrac{9}{24} - \dfrac{4}{24} = \dfrac{9 - 4}{24} = \dfrac{5}{24}$, and $\dfrac{2}{7} - \dfrac{5}{7} = \dfrac{2 - 5}{7} = \dfrac{-3}{7}$.

In the case of division, let us first consider a particular case, say, $\frac{5}{7} \div \frac{2}{3}$. By Definition 8.3.2, this is the number x (if there is one) such that

$$\frac{2}{3} \cdot x = \frac{5}{7}.$$

It is convenient to try to find an x of the form $3g/2h$, because then

$$\frac{2}{3} \cdot x = \frac{2}{3} \cdot \frac{3g}{2h} = \frac{2 \cdot 3 \cdot g}{3 \cdot 2 \cdot h} = \frac{g}{h}.$$

We can see that this will equal $\frac{5}{7}$ if we take $g = 5$ and $h = 7$; that is, if we take x to be

$$\frac{3 \cdot 5}{2 \cdot 7} = \frac{15}{14}.$$

Thus

$$\frac{2}{3} \cdot \frac{15}{14} = \frac{5}{7}$$

and hence, by the definition of division,

$$\frac{5}{7} \div \frac{2}{3} = \frac{5 \cdot 3}{7 \cdot 2} = \frac{15}{14}.$$

For the general case of division of one fraction by another, we will consider

$$\frac{a}{b} \div \frac{c}{d}.$$

By definition, this is the number x (if there is one) such that

$$\frac{c}{d} \cdot x = \frac{a}{b}.$$

It is convenient to try to find an x of the form dg/ch because then

$$\frac{c}{d} \cdot x = \frac{c}{d} \cdot \frac{dg}{ch} = \frac{cdg}{dch} = \frac{g}{h},$$

and we can see that this will equal a/b if we take $g = a$ and $h = b$; that is, if we take x to be ad/bc. Thus by the definition of division,

$$\frac{a}{b} \div \frac{c}{d} = \frac{ad}{bc} = \frac{a}{b} \times \frac{d}{c}.$$

We have proved the following theorem:

Theorem 8.3.3. *If $\dfrac{a}{b}$ and $\dfrac{c}{d}$ are fractions and $c \neq 0$, then*

$$\frac{a}{b} \div \frac{c}{d} = \frac{a}{b} \times \frac{d}{c} = \frac{ad}{bc}.$$

Note that if $c = 0$ then ad/bc is not a fraction. When $c \neq 0$ and $d \neq 0$, the fraction c/d is called the **reciprocal** of d/c. In particular, the reciprocal of an integer c that is not 0 is $1/c$. We can thus say that to divide by a fraction, multiply by the reciprocal of the fraction. For example,

$$\frac{7}{8} \div \frac{3}{4} = \frac{7}{8} \times \frac{4}{3} = \frac{7 \cdot 4}{8 \cdot 3} = \frac{7}{2 \cdot 3} = \frac{7}{6}.$$

We shall now see that the rational number a/b, which according to our notation is the ordered pair $[a, b]$ of $I \times N$, actually turns out to be $a \div b$. In fact, this is what we wanted to happen and is the motivation for our definitions of addition and multiplication in F (Definition 8.1.2).

Theorem 8.3.4. *For all $a, b \in I$ with $b \neq 0$, $\dfrac{a}{b} = a \div b$.*

Proof. Using $a = a/1$, $b = b/1$, and Theorem 8.3.3, we have

$$a \div b = \frac{a}{1} \div \frac{b}{1} = \frac{a}{1} \cdot \frac{1}{b} = \frac{a}{b} .$$

Thus the rational numbers can be considered to be the enlargement of the system of integers that makes division of integers always possible, as long as the divisor is not zero. That is, if a and b are integers and $b \neq 0$, in our enlarged system F, we can now find x such that $bx = a$, namely $x = a/b$.

Exercises

1. Use the *definitions* of subtraction and division in F (not Theorems 8.3.1 and 8.3.3) to prove that

   (a) $\dfrac{7}{3} - \dfrac{0}{5} = \dfrac{7}{3}$;

   (b) $\dfrac{2}{3} - \dfrac{5}{8} = \dfrac{1}{24}$;

   (c) $\dfrac{3}{4} \div \dfrac{15}{8} = \dfrac{2}{5}$;

   (d) $\dfrac{16}{12} \div \dfrac{56}{21} = \dfrac{1}{2}$.

   Find the following quotients and differences, expressing your results in lowest terms.

2. $\dfrac{-7}{12} - \left(\dfrac{3}{8} - \dfrac{1}{10} \right)$;

3. $\left(\dfrac{-7}{12} - \dfrac{3}{8} \right) - \dfrac{1}{10}$;

4. $\dfrac{2}{3} \div \left(\dfrac{3}{4} - \dfrac{7}{10} \right)$;

5. $\dfrac{15}{2} - \dfrac{26}{5}$;

6. $\dfrac{16}{15} \div \dfrac{2}{7}$;

7. $\left(\dfrac{1}{4} - \dfrac{-7}{6} \right) \div \dfrac{3}{8}$;

8. $\left(\dfrac{7}{16} \div \dfrac{3}{8} \right) \div \dfrac{5}{4}$;

9. $\dfrac{7}{16} \div \left(\dfrac{3}{8} \div \dfrac{5}{4} \right)$;

10. $\dfrac{21}{16} \div \left(\dfrac{-3}{14} - \dfrac{3}{4} \right)$.

11. Use Theorem 8.3.3 to prove that if c/d is in F and $c \neq 0$, then $0 \div c/d = 0$.

12. Prove that subtraction and division are not commutative in F by proving the following:
    (a) There exist numbers f and g in F such that $f - g \neq g - f$;
    (b) There exist numbers f and g in F such that $f \div g \neq g \div f$.

13. Show that there exist numbers a and b in I such that $\dfrac{1}{a} + \dfrac{1}{b} \neq \dfrac{1}{a+b}$.

14. Explain why each of the following equations follows directly from the definition of division.

(a) $g(f \div g) = f$;   (b) $(fh) \div h = f$.

15. Explain why each of the following equations follows directly from the definition of subtraction.

(a) $g + (f - g) = f$;   (b) $(f + h) - h = f$.

8.4 Some properties of addition, subtraction, multiplication, and division

We shall now establish some familiar properties of addition, subtraction, multiplication, and division.

Theorem 8.4.1. *If f, g, and h are in F, then:*

(1) $f + g \in F$ *(F is closed under addition);*
(2) $f + g = g + f$ *(Addition in F is commutative);*
(3) $(f + g) + h = f + (g + h)$ *(Addition in F is associative);*
(4) $f + 0 = 0 + f = f$ *(0 is the identity of addition);*
(5) $f - g \in F$ *(F is closed under subtraction);*
(6) $fg \in F$ *(F is closed under multiplication);*
(7) $fg = gf$ *(Multiplication in F is commutative);*
(8) $(fg)h = f(gh)$ *(Multiplication in F is associative);*
(9) $f \cdot 1 = 1 \cdot f = f$ *(1 is the identity of multiplication);*
(10) $f(g + h) = fg + fh, (g + h)f = gf + hf$ *(Multiplication is distributive on both sides over addition);*
(11) *If $g \neq 0$ then $f \div g \in F$ (F is closed under division, except by 0).*

Proof. We shall prove parts (1), (2), (4,) (8), and (11), leaving the proofs of the other parts as exercises. Let $f = a/b$, $g = c/d$, and $h = r/s$ where a, c, and r are integers and b, d, and s are nonzero integers.

(1) $f + g = \dfrac{a}{b} + \dfrac{c}{d} = \dfrac{ad + bc}{bd}$. But by the closure properties of addition

and multiplication in I, $ad + bc \in I$ and $bd \in I$. Also $bd \neq 0$, because $b \neq 0$ and $d \neq 0$ by the contrapositive of Theorem 6.7.2, part (2). Therefore

$$\frac{ad + bc}{bd} \in F,$$

so $f + g \in F$.

(2) By the definition of addition in F,

$$f + g = \frac{a}{b} + \frac{c}{d} = \frac{ad + bc}{bd} \qquad \text{and} \qquad g + f = \frac{c}{d} + \frac{a}{b} = \frac{cb + da}{db}$$

But by the commutative properties in I,

$$\frac{ad + bc}{bd} = \frac{cb + da}{db},$$

hence $f + g = g + f$.

(4) $f + 0 = \dfrac{a}{b} + \dfrac{0}{1} = \dfrac{a \cdot 1 + b \cdot 0}{b \cdot 1} = \dfrac{a + 0}{b} = \dfrac{a}{b} = f$, hence by the transitive property of equality in F, $f + 0 = f$. But by (2), $f + 0 = 0 + f$, and therefore $f + 0 = 0 + f = f$.

(8) By definition of multiplication and the associative property of multiplication in I,

$$(fg)h = \left(\frac{a}{b} \cdot \frac{c}{d}\right)\frac{r}{s} = \frac{ac}{bd} \cdot \frac{r}{s} = \frac{acr}{bds},$$

and similarly

$$f(gh) = \frac{a}{b}\left(\frac{c}{d} \cdot \frac{r}{s}\right) = \frac{a}{b} \cdot \frac{cr}{ds} = \frac{acr}{bds}.$$

Therefore $(fg)h = f(gh)$.

(11) If $g \neq 0$, we have $c/d \neq 0/1$. Therefore $c \cdot 1 \neq d \cdot 0 = 0$, so $c \neq 0$. Now

$$f \div g = \frac{a}{b} \div \frac{c}{d} = \frac{a}{b} \cdot \frac{d}{c} = \frac{ad}{bc}.$$

But since I is closed under multiplication, ad is in I, and since the nonzero integers are closed under multiplication by Theorem 6.7.2, $bc \neq 0$. Hence $f \div g = ad/bc$ is in F.

Similarly to the case of whole numbers in Section 4.4, it follows from the commutative and associative properties above that any sum or product is independent of the way it is associated and independent of the order of the terms. For example it follows that if f, g, h, k are in F. then

$$(f + g) + (h + k) = [(k + g) + f] + h$$

and

$$(fg \cdot h)k = (k \cdot hf)g = (kf)(gh).$$

Thus we can write $f + g + h + k$, $fghk$, and other expressions without ambiguity.

We now obtain two "cancellation properties" for rational numbers. Our proofs use corresponding cancellation properties of I indicated in Theorems 6.3.2 and 6.7.3.

Theorem 8.4.2. *Let f, g, and h be rational numbers. Then*
   (1) $f = g$ *if and only if* $f + h = g + h$;
   (2) *When* $h \neq 0$, $f = g$ *if and only if* $fh = gh$.

Proof. Since addition and multiplication are not ambiguous in F (Theorem 8.2.1), if $f = g$, then (since $h = h$) $f + h = g + h$ and $fh = gh$. We have thus proved the "only if" parts of the two double implications.

To prove the "if" parts, let $f = a/b$, $g = c/d$, and $h = r/s$ where a, b, c, d, r, s are in I and none of the denominators are zero.
   (1) If $f + h = g + h$, then

$$f + h = \frac{a}{b} + \frac{r}{s} = \frac{as + br}{bs} = g + h = \frac{c}{d} + \frac{r}{s} = \frac{cs + dr}{ds},$$

and so

$$\frac{as + br}{bs} = \frac{cs + dr}{ds}.$$

Hence, by definition of equality in F, $(as + br)ds = bs(cs + dr)$. Using a cancellation property of I, we can cancel the s to get $(as + br)d = b(cs + dr)$. Then, by the distributive property, we obtain

$$asd + brd = bcs + bdr.$$

Again using cancellation properties of I, we can cancel the brd and bdr to get $asd = bcs$, and then cancel the s to get $ad = bc$. Consequently, by definition of equality in F, we have $a/b = c/d$; that is, $f = g$. Therefore, if $f + h = g + h$, then $f = g$.
   (2) When $h \neq 0$ and $fh = gh$, we have

$$fh = \frac{a}{b} \cdot \frac{r}{s} = \frac{ar}{bs} = gh = \frac{c}{d} \cdot \frac{r}{s} = \frac{cr}{ds}.$$

Therefore $ar/bs = cr/ds$, and by definition of equality in F we get $ards = bscr$. Now since $h \neq 0$, $r \neq 0$ and since s is a denominator, $s \neq 0$. Thus by a cancellation property of I, we can cancel the r and s to get $ad = bc$. Then, by definition of equality, we have $f = g$. Therefore, when $h \neq 0$, if $fh = gh$ then $f = g$.

Notice that even when $h = 0$, if $f = g$ then $fh = gh$, but the converse is false, For example, if $x = y$, then $x \cdot 0 = y \cdot 0 = 0$; however, $5 \cdot 0 = 3 \cdot 0$ but $5 \neq 3$.

Theorem 8.4.2 has many uses. In particular, it can be used to help prove the next three theorems.

Theorem 8.4.3. *Subtraction and division (when the divisor is not zero) are unique in F.*

Proof. Let f and g be rational numbers and let x and x' be rational numbers such that $g + x = f$ and $g + x' = f$. Then $g + x = g + x'$ and, by Theorem 8.4.2, we can cancel the g to get $x = x'$; so subtraction is unique in F. Now if $g \neq 0$, and x and x' are rational numbers such that $gx = f$ and $gx' = f$, then $gx = gx'$. Since $g \neq 0$ we can cancel it, by Theorem 8.4.2, to get $x = x'$, and hence division is also unique in F.

Theorem 8.4.4. *Let f, g and h be rational numbers. Then*

(1) *$f = g$ if and only if $f - h = g - h$;*
(2) *When $h \neq 0$, $f = g$ if and only if $f \div h = g \div h$.*

Proof. Because of the way subtraction and division are defined in F in terms of addition and multiplication, it follows that subtraction and division are not ambiguous. Hence if $f = g$ then $f - h = g - h$, and if $h \neq 0$ then $f \div h = g \div h$. Thus we have proved the "only if" parts of the theorem and will now prove the "if" parts.

(1) If $f - h = g - h$, then by the "only if" part of Theorem 8.4.2, part (1), $(f - h) + h = (g - h) + h$. But $(f - h) + h = f$ and $(g - h) + h = g$ by definition of subtraction, so $f = g$.

(2) When $h \neq 0$ and $f \div h = g \div h$, then by the "only if" part of Theorem 8.4.2, part (2), $(f \div h)h = (g \div h)h$. But by definition of division, $(f \div h)h = f$ and $(g \div h)h = g$, so $f = g$.

Theorem 8.4.5. *Let f, g, and h be rational numbers. Then*
(1) *$f \cdot 0 = 0$;*
(2) *$fg = 0$ if and only if $f = 0$ or $g = 0$;*
(3) *$f(g - h) = fg - fh$ and $(g - h)f = gf - hf$.*

Proof.
(1) Let $f = a/b$ where a and b are in I and $b \neq 0$. Then

$$f \cdot 0 = \frac{a}{b} \cdot \frac{0}{1} = \frac{a \cdot 0}{b \cdot 1} = \frac{0}{b} = \frac{0}{1} = 0,$$

so $f \cdot 0 = 0$.

(2) If $f = 0$ or $g = 0$, then by (1) and the commutative property of multiplication, $fg = 0$. Conversely, if $fg = 0$ and $g \neq 0$, then since $0 \cdot g = 0$, we have $fg = 0 \cdot g$; by Theorem 8.4.2 we can cancel the g to get $f = 0$. We have shown that if $fg = 0$ and $g \neq 0$, then $f = 0$. Thus, if $fg = 0$ then $f = 0$ or $g = 0$.

(3) The proof is essentially the same as the proof of the corresponding theorem (Theorem 4.8.1) for whole numbers and thus will be omitted.

We next state two theorems that are similar to Theorems 6.7.1 and 6.7.4. Their proofs are essentially the same as those of Theorems 6.7.1 and 6.7.4 and will be omitted.

Theorem 8.4.6. *If f, g, and h are rational numbers, then*
(1) $(f + g) - h = f + (g - h) = (f - h) + g$;
(2) $(f - g) - h = (f - h) - g$.

Theorem 8.4.7. *If f, g, and h are rational numbers and $h \neq 0$, then*
(1) $(f + g) \div h = (f \div h) + (g \div h)$;
(2) $(f - g) \div h = (f \div h) - (g \div h)$.

That is, division is distributive on the right over addition and subtraction. For example, the reader can verify that

$$\left(\frac{3}{4} + \frac{2}{3}\right) \div \frac{4}{5} = \frac{3}{4} \div \frac{4}{5} + \frac{2}{3} \div \frac{4}{5}.$$

Division is, however, not distributive on the left over addition or subtraction.

Exercises

1. Prove part (9) of Theorem 8.4.1.   2. Prove part (7) of Theorem 8.4.1.
3. Prove part (6) of Theorem 8.4.1.   4. Prove part (3) of Theorem 8.4.1.
5. Use Theorem 8.3.1. to help prove part (5) of Theorem 8.4.1.
6. Prove part (10) of Theorem 8.4.1.
7. Prove that subtraction and division are not associative in F by proving the following:
   (a) There exist numbers f, g, and h in F such that $f - (g - h) \neq (f - g) - h$;
   (b) There exist numbers f, g, and h in F such that $f \div (g \div h) \neq (f \div g) \div h$.
8. Prove that division is not distributive on the left over addition or subtraction, by showing that there exist numbers f, g, and h in F such that
   (a) $h \div (f + g) \neq (h \div f) + (h \div g)$;
   (b) $h \div (f - g) \neq (h \div f) - (h \div g)$.

8.5   Positive and negative rational numbers

In this section we shall be considering positive and negative rational numbers as well as the concept of the negative of a rational number.

Definition 8.5.1. *A positive rational number is one that can be expressed as one whose numerator and denominator are positive integers.*

In Theorem 6.8.1 we showed that the positive integers are closed under addition and multiplication. The positive rational numbers have this property and are also closed under division, as our next theorem indicates.

Theorem 8.5.1. *The set of positive rational numbers is closed under addition, multiplication, and division.*

Proof. Let $f = a/b$ and $g = c/d$ be positive rational numbers where a, b, c, d are positive integers. Now

$$f + g = \frac{a}{b} + \frac{c}{d} = \frac{ad + bc}{bd} \; ; \; fg = \frac{ac}{bd} \; ; \; f \div g = \frac{a}{b} \cdot \frac{d}{c} = \frac{ad}{bc} \; .$$

But since the positive integers are closed under addition and multiplication, $ad + bc$, ac, bd, ad, and bc are positive integers, and hence $f + g$, fg, and $f \div g$ are positive rational numbers. Thus, the positive rational numbers are closed under addition, multiplication, and division.

Similarly to Definition 6.4.2, we define the negative of a rational number as follows.

Definition 8.5.2. *If f is a rational number, then the number x such that $f + x = 0$ is called the **negative** of f and is written as $-f$.*

For example,

$$\frac{2}{3} + \frac{-2}{3} = \frac{2 + (-2)}{3} = \frac{0}{3} = 0,$$

so $\dfrac{-2}{3}$ is the negative of $\dfrac{2}{3}$, that is,

$$-\frac{2}{3} = \frac{-2}{3}.$$

Similarly,

$$\frac{-2}{3} + \frac{2}{3} = \frac{(-2) + 2}{3} = \frac{0}{3} = 0,$$

so $\dfrac{2}{3}$ is the negative of $\dfrac{-2}{3}$. That is,

$$-\frac{-2}{3} = \frac{2}{3}.$$

More generally, we have the next theorem, which is not only an aid to simplifying expressions but also proves that for every rational number f, its negative $-f$ exists and is also a rational number.

Theorem 8.5.2. *If $\dfrac{a}{b}$ is any rational number, then*

$$\frac{-a}{b} = \frac{a}{-b} = -\frac{a}{b}.$$

Proof. By Theorem 8.1.2,

$$\frac{a}{-b} = \frac{(-1)a}{(-1)(-b)} = \frac{-a}{b}.$$

Also, by definition of $-a$ in I,

$$\frac{a}{b} + \frac{-a}{b} = \frac{a + (-a)}{b} = \frac{0}{b} = 0,$$

and hence

$$\frac{-a}{b} = -\frac{a}{b}.$$

Therefore

$$\frac{-a}{b} = \frac{a}{-b} = -\frac{a}{b}.$$

For example,

$$\frac{-3}{4} = \frac{3}{-4} = -\frac{3}{4},$$

and

$$-\frac{-5}{7} = \frac{-(-5)}{7} = \frac{5}{7} = \frac{-5}{-7}.$$

Theorem 8.5.3. *If f is a rational number, then $-(-f) = f$.*

Proof. By Definition 8.5.2, the negative $-(-f)$ of $-f$ is the number x such that $-f + x = 0$. Also by Definition 8.5.2, $f + (-f) = 0$. But by Theorem 8.4.1, addition is commutative in F, and therefore $(-f) + f = 0$; hence by definition of the negative of $-f$, $-(-f) = f$.
For example, $-(-\frac{4}{3}) = \frac{4}{3}$ and $-(-(-\frac{7}{9})) = -\frac{7}{9}$.

Definition 8.5.3. *The negative of a positive rational number is called a **negative rational number**.*

For example, $-\frac{19}{31}$ is a negative rational number.

Similarly to the integers, it should be realized that the negative of a rational number is not necessarily a negative rational number. For example, the negative $-(-\frac{4}{5})$ of $-\frac{4}{5}$ is the positive rational number $\frac{4}{5}$. In general we have the following theorem.

Theorem 8.5.4. *Let f be a rational number. Then*
  (1) *If f is a **positive** rational number then $-f$ is a **negative** rational number;*
  (2) *If $f = 0$ then $-f = 0$;*
  (3) *If f is a **negative** rational number, then $-f$ is a **positive** rational number.*

  Proof.
  (1) This is just a restatement of the definition of a negative rational number.
  (2) Since $0 + 0 = 0$, $-0 = 0$ by the definition of negative.
  (3) If f is a negative rational number, then by definition, $f = -g$ for a positive rational number g; by Theorem 8.5.3, $-f = -(-g) = g$, and hence $-f$ is the positive number g.

Theorem 8.5.5. *If f is a rational number, then one and only one of the following holds.*
  (1) *f is positive;*   (2) *$f = 0$;*   (3) *f is negative.*

  Proof. Let $f = a/b$ where a and b are integers and $b \neq 0$. If $a = 0$ then $f = 0/b = 0$. Suppose that $a \neq 0$. Then if a and b are both positive, f is positive by definition. If both a and b are negative, then

$$f = \frac{a}{b} = \frac{(-1)a}{(-1)b} = \frac{-a}{-b}$$

where both $-a$ and $-b$ are positive; again f is positive. If a is positive and b is negative, let $b = -h$ where h is positive; then by Theorem 8.5.2,

$$f = \frac{a}{b} = \frac{a}{-h} = -\frac{a}{h}$$

and by definition, f is negative. Similarly, if a is negative and b is positive, let $a = -h$ where h is positive; then

$$f = \frac{a}{b} = \frac{-h}{b} = -\frac{h}{b},$$

and hence f is negative. We have now shown that at least one of the three properties holds.

If $f = a/b$ is positive or negative then $a \neq 0$, and if $a/b = 0/1$ then $a = b \cdot 0 = 0$, a contradiction. Therefore f cannot be both positive, or negative, and zero. If f were both positive and negative, then, by definition, $f = a/b$ when both a and b are positive and $f = -c/d$ when c and d are both positive. Thus

$$\frac{a}{b} = -\frac{c}{d} = \frac{-c}{d}$$

and hence by definition of equality in F, $ad = b(-c) = -(bc)$. But ad is a positive integer and $-(bc)$ is a negative integer, and an integer cannot be both positive and negative by Theorem 6.4.2. Hence, at most one of the three properties holds, and the theorem is proved.

By Theorem 8.5.5, every rational number is either positive, negative, or zero. Thus every rational number can be expressed as a/b where a and b are both positive integers, as $-a/b$ where a and b are both positive integers, or as 0. One of these will usually be the most convenient way of expressing a rational number.

Exercises

1. Write the negative of each of the following rational numbers:

(a) $\frac{7}{3}$; (b) -6; (c) $-\frac{5}{2}$; (d) 0; (e) -0; (f) $\frac{0}{7}$;

(g) $-\frac{2}{11}$; (h) $\frac{-3}{7}$; (i) $\frac{2}{-11}$; (j) $\frac{-3}{-4}$; (k) $\frac{0}{-3}$; (l) $-\frac{-3}{-5}$.

2. Prove that the positive rational numbers are not closed under subtraction.

8.6 Inequalities

We shall now extend the definition of $<$ and $>$ to include relations between rational numbers.

Definition 8.6.1. *Let f and g be rational numbers; then $f < g$ and $g > f$ if and only if there exists a positive rational number h such that $f + h = g$.*

Let us see what this means in terms of the numerators and denominators. To do this, let $f = a/b$ and $g = c/d$ where a, b, c, d are in I and b, d are in P

(the set of positive integers), and suppose that $f < g$. Then there is a positive fraction p/q, where p and q are in P, such that

$$\frac{a}{b} + \frac{p}{q} = \frac{c}{d}.$$

Then

$$\frac{aq + bp}{bq} = \frac{c}{d},$$

and hence $aqd + bpd = cbq$. Now, since b, p, and d are positive integers and P is closed under multiplication, by the definition of $<$ in I, we have $aqd < cbq$. By Theorem 6.8.4, we can cancel the q to get

$$ad < bc.$$

Therefore if $a/b < c/d$, then $ad < bc$. Does the converse also hold? That is, if $ad < bc$ is it true that $a/b < c/d$? Yes, for if $ad < bc$ then there is a positive integer p such that $ad + p = bc$. Hence

$$\frac{(ad + p)}{bd} = \frac{bc}{bd}.$$

Therefore

$$\frac{ad}{bd} + \frac{p}{bd} = \frac{bc}{bd},$$

and we have

$$\frac{a}{b} + \frac{p}{bd} = \frac{c}{d}.$$

Thus, since p and bd are positive, $a/b < c/d$. We have now proved part (1) of the following theorem. The proof of part (2) is left as an exercise.

Theorem 8.6.1. *Let a, b, c, and d be in I where b and d are positive. Then*

(1) $\dfrac{a}{b} < \dfrac{c}{d}$ *if and only if $ad < bc$;*

(2) $\dfrac{a}{b} > \dfrac{c}{d}$ *if and only if $ad > bc$.*

For example, $4/7 < 5/8$ because $4 \cdot 8 = 32 < 7 \cdot 5 = 35$, and $-3/4 < -1/2$ because $(-3)2 = -6 < 4(-1) = -4$. Be careful to realize that Theorem 8.6.1 requires both denominators to be positive. Actually it still holds if both denominators are negative but fails if one denominator is positive and the other is negative. For example, $4/-2 < 6/2$ but $(4)(2) > (-2)(6)$.

Notice that according to the definition of $>$, a rational number $f > 0$ if and only if there is a positive integer h such that $f = 0 + h = h$. Thus a rational number if greater than zero if and only if it is positive. We can now obtain the trichotomy property for rational numbers.

Theorem 8.6.2. *If f and g are rational numbers, then one and only one of the following must hold:*

(1) $f < g$;   (2) $f = g$;   (3) $f > g$.

Proof. This follows directly from the trichotomy property in I, as we shall now see. Let $f = a/b$ and $g = c/d$ where a, b, c, d are in I and b and d are positive. Then by the trichotomy property in I, one and only one of the following holds:

(1) $ad < bc$,   (2) $ad = bc$,   (3) $ad > bc$.

Hence by Theorem 8.6.1, one and only one of the following must hold:

(1) $f < g$,   (2) $f = g$,   (3) $f > g$.

We shall now obtain some properties of inequalities of rational numbers that are similar to the corresponding properties of integers in Theorem 6.8.4.

Theorem 8.6.3. *Let f, g and h be rational numbers, Then*

(1) *If $f < g$ and $g < h$, then $f < h$;*
(2) *$f < g$ if and only if $f + h < g + h$;*
(3) *$f < g$ if and only if $f - h < g - h$;*
(4) *When $h > 0$, $f < g$ if and only if $fh < gh$;*
(5) *When $h < 0$, $f < g$ if and only if $fh > gh$.*

Proof. The proofs of parts (1), (2), and (3) are essentially the same as the proofs of the corresponding parts of Theorem 6.8.4.

(4) Assume that $h > 0$. The proof that if $f < g$ then $fh < gh$ is left as an exercise, and we shall only prove that if $fh < gh$ then $f < g$. Since $fh < gh$, there exists a positive rational number k such that $fh + k = gh$. Therefore, because division on the right is distributive over addition (Theorem 8.4.7),

$$(fh \div h) + (k \div h) = gh \div h.$$

Hence $f + (k \div h) = g$. But $k \div h$ is positive by Theorem 8.5.1, and so $f < g$.

(5) The proof of this part is left as an exercise.

Exercises

1. Arrange the following numbers in proper order from left to right, using the symbol "$<$" or the symbol "$=$" between successive numbers as you write them.

$$-\frac{17}{35}, \ -\frac{19}{38}, \ \frac{4}{7}, \ \frac{5}{9}, \ \frac{17}{35}, \ \frac{19}{38}, \ \frac{7}{13}, \ \frac{8}{17}, \ \frac{21}{43}, \ \frac{1}{2}, \ \frac{1}{4}, \ \frac{2}{3}, \ -\frac{7}{13}, \ 0$$

2. Prove the property of Theorem 8.6.3, part (4), that was left as an exercise.

3. Prove that if f and g are both rational numbers such that $f < g$, then

$$2f < f + g < 2g \text{ and hence } f < \tfrac{1}{2}(f + g) < g.$$

4. Prove part (5) of Theorem 8.6.3.

5. Prove part (2) of Theorem 8.6.1 from part (1).

8.7 Proper and improper fractions

Let a be a positive integer or zero, and let b be a positive integer. Then a/b is a positive fraction or zero and $-a/b$ is a negative fraction or zero, and every rational number can be expressed in one of these two forms.

Definition 8.7.1. *Let a be a positive integer or zero and let b be a positive integer. The fractions a/b and $-a/b$ are called **improper fractions** when $a \geq b$ and **proper fractions** when $a < b$.*

For example, $\frac{5}{3}$, $-\frac{5}{3}$, $\frac{8}{8}$, and $-\frac{11}{4}$ are improper fractions and $\frac{2}{3}$, $\frac{3}{4}$, $-\frac{7}{19}$, and $-\frac{6}{9}$ are proper fractions.

Definition 8.7.2. *A **mixed number** is an expression for a rational number as the sum of a positive integer and a positive proper fraction, or the negative of such a sum.*

For example, $5 + \frac{3}{5}$ and $-(4 + \frac{1}{3})$ are mixed numbers, and for convenience we write $5 + \frac{3}{5}$ as $5\frac{3}{5}$ and $-(4 + \frac{1}{3})$ as $-4\frac{1}{3}$. We express $\frac{23}{3}$ and $-\frac{23}{3}$ as the mixed numbers $7\frac{2}{3}$ and $-7\frac{2}{3}$ respectively, as follows. Use the division algorithm to write

$$23 = 3(7) + 2 \qquad \text{where} \qquad 0 \leq 2 < 3.$$

Then

$$\frac{23}{3} = (3 \cdot 7 + 2) \div 3 = \frac{3 \cdot 7}{3} + \frac{2}{3} = 7 + \frac{2}{3}.$$

Thus $\frac{23}{3} = 7 + \frac{2}{3}$ and $-\frac{23}{3} = -(7 + \frac{2}{3})$, and for convenience we write $7 + \frac{2}{3}$ as $7\frac{2}{3}$ and $-(7 + \frac{2}{3})$ as $-7\frac{2}{3}$.

In general, for any two positive integers a and b for which $a > b$, in order to express the improper fractions a/b or $-a/b$ as mixed numbers, we first use the **division algorithm** of Section 4.9 to obtain integers q and r such that

$$a = bq + r, \qquad \text{where} \qquad 0 \leq r < b.$$

Then

$$\frac{a}{b} = (bq + r) \div b = q + \frac{r}{b}.$$

Thus $a/b = q + r/b$ and $-a/b = -(q + r/b)$ where r/b is either 0 or a proper fraction, since $0 \le r < b$. It is customary to omit the $+$ sign in a mixed number, as we did when we wrote $7 + \frac{2}{3}$ as $7\frac{2}{3}$.

Exercises

Express each of the following improper fractions as a mixed number.

1. $\dfrac{68}{4}$; **2.** $\dfrac{249}{17}$; **3.** $-\dfrac{92}{3}$; **4.** $-\dfrac{1287}{5}$; **5.** $\dfrac{371}{29}$; **6.** $-\dfrac{87}{31}$.

Perform the indicated operations, and write each result as a mixed number or a proper fraction.

7. $8\frac{3}{5} + 7\frac{2}{3} - 3\frac{1}{2}$; **8.** $8\frac{3}{8} \times 4\frac{1}{3}$;

9. $8\frac{2}{7} - 5\frac{1}{3}$; **10.** $(4\frac{3}{4} - 3\frac{1}{6}) \times 2\frac{2}{3}$.

8.8   Rational numbers on the number line

Recall that we put the integers in 1–1 correspondence with a subset of the points on a line by labeling one point 0, taking a convenient unit of measure, and then, for each $n \in I$ that is positive or zero, assigning the label n to the point that is n units to the right of 0. For each $-p$ of I that is negative (and hence p is positive) we assigned the label $-p$ to the point that is p units to the left of the origin.

This gives us the fractions with denominator 1 as labels for appropriate points. To obtain points for the fractions with denominator 2, take a unit of measure that fits exactly 2 times on the segment from 0 to 1. Then assign

the label 1/2 to the point that is one such unit to the right of 0, the label
2/2 = 1 to the point that is two of these units to the right of 0, the label 3/2 to
the point that is three of these units to the right of 0, etc.; the label −1/2 to
the point that is one of these units to the left of the origin, the label −2/2 = − 1
to the point that is 2 such units to the left of the origin, etc. In general, to
label points with denominator $d > 0$, take a unit of measure that fits exactly
d times on the segment from 0 to 1. Then give the label $1/d$ to the point that is
one of these units to the right of 0, the label $2/d$ to the point that is 2 of
these units to the right of 0, etc.; that is, for each positive integer h, give the
label h/d to the point h such units to the right of the origin and the label
$−h/d$ to the point that is h such units to the left of the origin. Thus every
rational number is the the label of a unique point on the number line, and,
for positive integers a and b, a point labeled a/b is a/b units to the right of the
origin, and a point labeled $−a/b$ is a/b units to the left of the origin.

Some labeled points between −1 and 1 are illustrated as follows:

When this technique is used, each such labeled point will have many
labels, but the rational numbers used for these labels will all be equal, and
equal rational numbers will always be labels for the same point. That is, for
rational numbers f and g, $f = g$ if and only if the points f and g are the same
distance and direction from the origin. Thus, $f = g$ if and only if the points f
and g are the same point.

As another consequence of the fact that, for a positive number k, the
point k is k units to the right of the origin, we see that at least when the numbers involved are positive, $f + h = g$ if and only if g is h units to the right of f
on the number line.

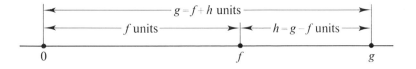

Figure 8.8.1

Now $f < g$ if and only if there exists a positive number h such that
$f + h = g$, and by definition of subtraction, $f + h = g$ if and only if f is to the
left of g on the number line and the distance between f and g is $g − f$ units,
as shown in Figure 8.8.1. Similarly $f > g$ if and only if f is to the right of g on
the number line and the distance between f and g is $f − g$ units.

Actually the additional fact that a negative number is the appropriate distance to the left of the origin on the number line can be used to show that the statements in the preceding paragraph hold, even if either f or g or both f and g are not positive. The details involve an exhaustive consideration of cases involving all the possible combinations of positive, negative, and zero values f and g can have, and we will not take the space to consider these details. It is worthwhile to look at a numerical example, however, and thus we will now consider one, with $f = -\frac{1}{2}$, $g = \frac{5}{2}$, and $h = \frac{6}{2} = 3$.

Example: $-\dfrac{1}{2} < \dfrac{5}{2}$ and $-\dfrac{1}{2} + \dfrac{6}{2} = \dfrac{-1}{2} + \dfrac{6}{2} = \dfrac{5}{2}$; that is, $-\dfrac{1}{2} + \dfrac{6}{2} = \dfrac{5}{2}$.

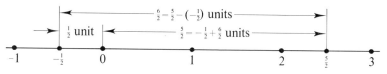

Figure 8.8.2

Thus, by the definition of subtraction, since

$$-\frac{1}{2} + \frac{6}{2} = \frac{5}{2}, \qquad \frac{6}{2} = \frac{5}{2} - \left(-\frac{1}{2}\right).$$

The point $-\frac{1}{2}$ is $\frac{6}{2} = 3$ units to the left of $\frac{5}{2}$, and the distance between $-\frac{1}{2}$ and $\frac{5}{2}$ is $\frac{5}{2} - (-\frac{1}{2}) = \frac{6}{2} = 3$ units, as shown in Figure 8.8.2.

It also follows that we can add, multiply, subtract, and divide rational numbers on the number line as we did in Section 6.5:

1. *To add $f + g$, go g units to the right of f.*
2. *To subtract $f - g$, go g units to the left of f.*
3. *To multiply fg, go f steps, of g units each, to the right of the origin.*
4. *To divide $f \div g$, determine how many g unit steps to the right are needed to go from the origin to f.*

We must treat negatives as in Section 6.5, however. For example, to go f steps of $-2\frac{1}{2}$ units to the right of the origin means to go f steps of $2\frac{1}{2}$ units to the left of the origin. Also to go -5 steps of g units to the right of the origin means to go 5 steps of g units to the left of the origin, and to determine how many $-2\frac{1}{2}$-unit steps to the right are needed to go from the origin to f, we determine how many $2\frac{1}{2}$-unit steps to the left are needed to go from the origin to f.

Let us consider the following more specific examples:

(1) To add $6 + (-2\frac{1}{2})$, go $2\frac{1}{2}$ units to the left of 6 to arrive at $3\frac{1}{2}$. Thus $6 + (-2\frac{1}{2}) = 3\frac{1}{2}$.

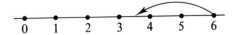

(2) To subtract $5 - (-2\frac{1}{2})$, go $2\frac{1}{2}$ units to the right of 5 to arrive at $7\frac{1}{2}$. Thus $5 - (-2\frac{1}{2}) = 7\frac{1}{2}$.

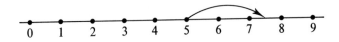

(3) To multiply $5(-2\frac{1}{2})$, go 5 steps of $2\frac{1}{2}$ units to the left of the origin, to arrive at $-12\frac{1}{2}$. Thus $5(-2\frac{1}{2}) = -12\frac{1}{2}$.

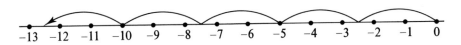

(4) To divide $5 \div (-2\frac{1}{2})$, determine how many $2\frac{1}{2}$-unit steps to the left are needed to go from the origin to 5. Since it takes two $2\frac{1}{2}$-unit steps to the right to go from the origin to 5, it takes minus two $2\frac{1}{2}$-unit steps to the left to go from the origin to 5. Thus $5 \div (-2\frac{1}{2}) = -2$.

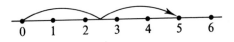

(5) To add $(-5) + 2\frac{1}{2}$, go $2\frac{1}{2}$ units to the right of -5 to arrive at $-2\frac{1}{2}$. Thus $(-5) + 2\frac{1}{2} = -2\frac{1}{2}$.

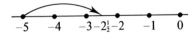

(6) To subtract $(-5) - 2\frac{1}{2}$, go $2\frac{1}{2}$ units to the left of -5 to arrive at $-7\frac{1}{2}$. Thus $(-5) - 2\frac{1}{2} = -7\frac{1}{2}$.

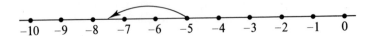

(7) To multiply $(-5)(2\frac{1}{2})$, go 5 steps of $2\frac{1}{2}$ units to the left of the origin to arrive at $-12\frac{1}{2}$. Thus $(-5)(2\frac{1}{2}) = -12\frac{1}{2}$.

(8) To divide $(-5) \div 2\frac{1}{2}$, determine how many $2\frac{1}{2}$-unit steps to the right are needed to go from the origin to -5. Since it takes two $2\frac{1}{2}$-unit steps to the left to go from the origin to -5, it takes minus two $2\frac{1}{2}$-unit steps to the right to go from the origin to -5. Therefore $(-5) \div 2\frac{1}{2} = -2$.

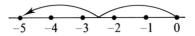

(9) To add $(-5) + (-2\frac{1}{2})$, go $2\frac{1}{2}$ units to the left of -5 to arrive at $-7\frac{1}{2}$. Thus $(-5) + (-2\frac{1}{2}) = -7\frac{1}{2}$.

(10) To subtract $(-5) - (-2\frac{1}{2})$, go $2\frac{1}{2}$ units to the right of -5 to arrive at $-2\frac{1}{2}$. Thus $(-5) - (-2\frac{1}{2}) = -2\frac{1}{2}$.

(11) To multiply $(-5)(-2\frac{1}{2})$, go -5 steps of $2\frac{1}{2}$ units to the left of the origin. That is, go 5 steps of $2\frac{1}{2}$ units to the right of the origin to arrive at $12\frac{1}{2}$. Thus $(-5)(-2\frac{1}{2}) = 12\frac{1}{2}$.

(12) To divide $(-5) \div (-2\frac{1}{2})$, determine how many $2\frac{1}{2}$-unit steps to the left are needed to go from the origin to -5. We see that 2 such steps are needed, so $(-5) \div (-2\frac{1}{2}) = 2$.

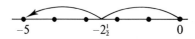

Note that in examples (3), (4), (7), and (8) where only one of the numbers in the product or quotient is negative, the result is negative, but in examples (11) and (12) where both numbers in the product or quotient are negative, the result is positive. We will see in the next section that positive and negative rational numbers always combine this way and hence that the results obtained in examples (1) to (12) above are correct.

Exercises

1. Draw a number line with 1 inch units and locate each of the following points on it:
   $0, 1, 2, 3, -1, -2, -3, -\frac{3}{4}, -\frac{5}{2}, \frac{5}{2}, -\frac{1}{4}, \frac{1}{4}, -3\frac{1}{2}$.

2. Write out a word description of how one locates each of the following points on a number line:
   (a) $\frac{2}{3}$;　　(b) $\frac{7}{3}$;　　(c) $\frac{17}{5}$;　　(d) $\frac{92}{37}$;　　(e) $-\frac{4}{3}$;　　(f) $-\frac{11}{17}$.

3. Describe in words, using the phrases "to the left of" or "to the right of," or "between," the set of all points x on the number line such that:
   (a) $x < \frac{7}{8}$;　　　(b) $x > -\frac{1}{2}$;　　　　(c) $-\frac{3}{4} < x < 2$;　　(d) $\frac{7}{16} > x > \frac{1}{8}$;
   (e) $x < 2$ and $x > 3$;　(f) $x < 2$ and $x < 1$;　　　(g) $x < -2$ or $x > 3$.
   Simplify your descriptions for (e) and (f).

4. What is the distance between each of the following pairs of points?
   (a) 2 and 5;　　(b) $\frac{1}{2}$ and $\frac{3}{7}$;　　(c) $\frac{27}{5}$ and $\frac{17}{13}$;　　(d) $\frac{24}{19}$ and 6.

5. Find each of the following mechanically on the number line by drawing appropriate figures:
   (a) $(-7) - 2$;　　(b) $\frac{7}{3} - (-\frac{2}{3})$;　　(c) $3(-\frac{2}{3})$;　　(d) $6 \div (-2)$;
   (e) $(-2)(-3)$;　　(f) $(-5) - (-3)$;　(g) $(-3) \div (-\frac{1}{2})$;　(h) $3\frac{1}{2} + (-\frac{3}{2})$;
   (i)　$-\frac{7}{8} - \frac{1}{2}$;　　(j) $(-3)(-\frac{1}{2})$;　　(k) $(-\frac{2}{3}) \div (-\frac{1}{3})$;　(l) $(-\frac{2}{3}) \div (\frac{1}{3})$.

8.9　Calculations with rational numbers

When combinations of positive and negative rational numbers are added, subtracted, multiplied, or divided, they behave in the same sort of way that positive and negative integers do as indicated in Theorem 6.6.1. In particular, the product or quotient of two negative rationals is positive, and when one rational is positive and the other negative, their product and quotient are negative. Thus products and quotients involving negative rationals can be found by performing the operations with positive numbers and then putting a minus sign in the appropriate place if necessary. Similarly, sums and differences involving negative rationals can be found by using suitable operations on positive rationals and appending a minus sign when necessary. The following theorem states the appropriate information:

Theorem 8.9.1. *Let a and c be non-negative integers and let b and d be positive integers. Then*

(1) $\left(-\dfrac{c}{d}\right) + \dfrac{a}{b} = \dfrac{a}{b} + \left(-\dfrac{c}{d}\right) = \dfrac{a}{b} - \dfrac{c}{d} = -\left(\dfrac{c}{d} - \dfrac{a}{b}\right);$

(2) $\left(-\dfrac{a}{b}\right) + \left(-\dfrac{c}{d}\right) = \left(-\dfrac{a}{b}\right) - \dfrac{c}{d} = \left(-\dfrac{c}{d}\right) - \dfrac{a}{b} = -\left(\dfrac{a}{b} + \dfrac{c}{d}\right);$

(3) $\left(-\dfrac{a}{b}\right) - \left(-\dfrac{c}{d}\right) = \left(-\dfrac{a}{b}\right) + \dfrac{c}{d} = \dfrac{c}{d} - \dfrac{a}{b} = -\left(\dfrac{a}{b} - \dfrac{c}{d}\right);$

(4) $\dfrac{a}{b} - \left(-\dfrac{c}{d}\right) = \dfrac{a}{b} + \dfrac{c}{d};$

(5) $\left(-\dfrac{a}{b}\right)\dfrac{c}{d} = \dfrac{a}{b}\left(-\dfrac{c}{d}\right) = -\left(\dfrac{a}{b} \cdot \dfrac{c}{d}\right);$

(6) $\left(-\dfrac{a}{b}\right)\left(-\dfrac{c}{d}\right) = \dfrac{a}{b} \cdot \dfrac{c}{d};$

(7) *If $c \neq 0$,* $\left(-\dfrac{a}{b}\right) \div \dfrac{c}{d} = \dfrac{a}{b} \div \left(-\dfrac{c}{d}\right) = -\left(\dfrac{a}{b} \div \dfrac{c}{d}\right);$

(8) *If $c \neq 0$,* $\left(-\dfrac{a}{b}\right) \div \left(-\dfrac{c}{d}\right) = \dfrac{a}{b} \div \dfrac{c}{d}.$

Proof. Throughout the proof we shall use Theorem 8.5.2 and Theorem 6.6.1 but, for brevity, will not refer to them. The reader should notice when they are used, however, to understand the proofs.

(1) By the commutative property of addition in F,

$$\frac{a}{b} + \left(-\frac{c}{d}\right) = \left(-\frac{c}{d}\right) + \frac{a}{b},$$

and

$$\left(-\frac{c}{d}\right) + \frac{a}{b} = \frac{-c}{d} + \frac{a}{b} = \frac{(-c)b + da}{db} = \frac{-cb + da}{bd} = \frac{ad - bc}{bd} = \frac{a}{b} - \frac{c}{d}.$$

Also,

$$\frac{a}{b} - \frac{c}{d} = \frac{ad - bc}{bd} = \frac{-(bc - ad)}{bd} = -\frac{bc - ad}{bd} = -\left(\frac{c}{d} - \frac{a}{b}\right),$$

and we have proved (1).

(2) $\left(-\dfrac{a}{b}\right) + \left(-\dfrac{c}{d}\right) = \dfrac{-a}{b} + \dfrac{c}{-d} = \dfrac{(-a)(-d) + bc}{b(-d)} = \dfrac{ad + bc}{-bd}$

$$= -\dfrac{ad + bc}{db} = -\left(\dfrac{a}{b} + \dfrac{c}{d}\right).$$

Also,

$$\left(-\dfrac{a}{b}\right) - \dfrac{c}{d} = \dfrac{-a}{b} - \dfrac{c}{d} = \dfrac{-ad - bc}{bd} = \dfrac{-(ad + bc)}{bd} = -\left(\dfrac{a}{b} + \dfrac{c}{d}\right).$$

It is left as an exercise to prove similarly that $\left(-\dfrac{c}{d}\right) - \dfrac{a}{b} = -\left(\dfrac{a}{b} + \dfrac{c}{d}\right).$

(3) $\left(-\dfrac{a}{b}\right) - \left(-\dfrac{c}{d}\right) = \dfrac{-a}{b} - \dfrac{c}{-d} = \dfrac{(-a)(-d) - bc}{b(-d)} = \dfrac{ad - bc}{-bd}$

$$= \dfrac{bc - ad}{bd} = \dfrac{c}{d} - \dfrac{a}{b}.$$

By part (1) (with a/b and c/d intechanged), however,

$$\left(-\dfrac{a}{b}\right) + \dfrac{c}{d} = \dfrac{c}{d} - \dfrac{a}{b} = -\left(\dfrac{a}{b} - \dfrac{c}{d}\right),$$

and we have proved (3).

(4) $\dfrac{a}{b} - \left(-\dfrac{c}{d}\right) = \dfrac{a}{b} - \dfrac{c}{-d} = \dfrac{a(-d) - bc}{b(-d)} = \dfrac{-ad - bc}{-bd}$

$$= \dfrac{ad + bc}{bd} = \dfrac{a}{b} + \dfrac{c}{d}.$$

(5) $\left(-\dfrac{a}{b}\right)\left(\dfrac{c}{d}\right) = \dfrac{-a}{b} \cdot \dfrac{c}{d} = \dfrac{(-a)c}{bd} = \dfrac{-ac}{bd} = -\dfrac{ac}{bd} = -\left(\dfrac{a}{b} \cdot \dfrac{c}{d}\right).$

Similarly, $\left(\dfrac{a}{b}\right)\left(-\dfrac{c}{d}\right) = -\left(\dfrac{a}{b} \cdot \dfrac{c}{d}\right).$

(6) $\left(-\dfrac{a}{b}\right)\left(-\dfrac{c}{d}\right) = \dfrac{-a}{b} \cdot \dfrac{-c}{d} = \dfrac{(-a)(-c)}{bd} = \dfrac{ac}{bd} = \dfrac{a}{b} \cdot \dfrac{c}{d}.$

(7) $\left(-\dfrac{a}{b}\right) \div \dfrac{c}{d} = \left(-\dfrac{a}{b}\right) \cdot \dfrac{d}{c} = -\left(\dfrac{a}{b} \cdot \dfrac{d}{c}\right)$ by (5), but $-\left(\dfrac{a}{b} \cdot \dfrac{d}{c}\right)$

$$= -\left(\dfrac{a}{b} \div \dfrac{c}{d}\right), \text{ so } \left(-\dfrac{a}{b}\right) \div \dfrac{c}{d} = -\left(\dfrac{a}{b} \div \dfrac{c}{d}\right).$$

Similarly, $\dfrac{a}{b} \div \left(-\dfrac{c}{d}\right) = -\left(\dfrac{a}{b} \div \dfrac{c}{d}\right)$.

$$(8) \quad \left(-\dfrac{a}{b}\right) \div \left(-\dfrac{c}{d}\right) = \left(\dfrac{-a}{b}\right) \div \left(\dfrac{c}{-d}\right) = \dfrac{-a}{b} \cdot \dfrac{-d}{c} = \dfrac{(-a)(-d)}{bc}$$

$$= \dfrac{ad}{bc} = \dfrac{a}{b} \div \dfrac{c}{d},$$

Examples:

(1) $-\frac{7}{8} + \frac{3}{4} = \frac{3}{4} - \frac{7}{8} = -(\frac{7}{8} - \frac{3}{4}) = -(\frac{7}{8} - \frac{6}{8}) = -\frac{1}{8}.$

(2) $-11 + 6 = -(11 - 6) = -5.$

(3) $(-\frac{1}{3}) - \frac{1}{2} = -(\frac{1}{3} + \frac{1}{2}) = -\frac{5}{6}; \quad (-\frac{1}{3}) + (-\frac{1}{2}) = -(\frac{1}{3} + \frac{1}{2}) = -\frac{5}{6}.$

(4) $(-2) - (-3) = -2 + 3 = 3 - 2 = 1;$

(5) $(-7) - (-2) = -7 + 2 = -(7 - 2) = -5.$

(6) $\frac{2}{3} - (-\frac{1}{3}) = \frac{2}{3} + \frac{1}{3} = \frac{3}{3} = 1.$ (7) $(-\frac{3}{4})(\frac{2}{3}) = -(\frac{3}{4} \cdot \frac{2}{3}) = -\frac{1}{2}.$

(8) $2(-6) = -(2 \cdot 6) = -12.$ (9) $(-\frac{1}{2})(-4) = \frac{1}{2}(4) = 2.$

(10) $(-7) \div (-3) = 7 \div 3 = \frac{7}{3}.$

(11) $(-\frac{2}{5}) \div (-\frac{1}{2}) = \frac{2}{5} \div \frac{1}{2} = \frac{2}{5} \cdot \frac{2}{1} = \frac{4}{5}.$

When f and g are rational numbers, it is sometimes a convenience to write f/g for $f \div g$. This is consistent with our previous notation a/b where a and b are integers, because we found that $a/b = a \div b$. In this notation, parts (7) and (8) of Theorem 8.9.1 can be restated as follows:

Theorem 8.9.2. *If f and g are rational numbers and $g \neq 0$, then*

$$\dfrac{-f}{g} = \dfrac{f}{-g} = -\dfrac{f}{g} \quad \text{and} \quad \dfrac{-f}{-g} = \dfrac{f}{g}.$$

Examples:

(1) $\dfrac{-7}{6} = \dfrac{7}{-6} = -\dfrac{7}{6}.$

(2) $\dfrac{-\frac{2}{3}}{-\frac{1}{6}} = \dfrac{\frac{2}{3}}{\frac{1}{6}} = \left(\dfrac{2}{3} \cdot \dfrac{6}{1}\right) = \dfrac{4}{1} = 4.$

(3) $\dfrac{-\frac{2}{3}}{\frac{3}{4}} = \dfrac{\frac{2}{3}}{-\frac{3}{4}} = -\dfrac{\frac{2}{3}}{\frac{3}{4}} = -\left(\dfrac{2}{3} \cdot \dfrac{4}{3}\right) = -\dfrac{8}{9}.$

Exercises

1. To complete the proof of part (5) of Theorem 8.9.1, prove that

$$\left(\dfrac{a}{b}\right)\left(-\dfrac{c}{d}\right) = -\left(\dfrac{a}{b} \cdot \dfrac{c}{d}\right)$$

2. Complete the proof of part (7) of Theorem 8.9.1 by showing that

$$\frac{a}{b} \div \left(-\frac{c}{d}\right) = -\left(\frac{a}{b} \div \frac{c}{d}\right).$$

Perform the following indicated operations:

3. $-\frac{2}{3} + \frac{3}{2}$.   **4.** $\frac{3}{4} + \left(-\frac{1}{2}\right)$.

5. $\left(-\frac{4}{5}\right) + \left(-\frac{2}{3}\right)$.   **6.** $\left(-\frac{3}{7}\right) - \left(-\frac{7}{2}\right)$.

7. $\left(-\frac{2}{5}\right) - \frac{7}{8}$.   **8.** $\left(-\frac{3}{4}\right)\left(-\frac{8}{9}\right)$.

9. $\left(-\frac{4}{15}\right) \div \left(-\frac{10}{21}\right)$.   **10.** $\left(-\frac{5}{8}\right)\left(\frac{4}{15}\right)$.

11. $\frac{5}{8}\left(-\frac{4}{15}\right)$.   **12.** $\left(-\frac{3}{4}\right) \div \frac{5}{8}$.

13. $\frac{3}{2} - \left(-\frac{5}{4}\right)$.   **14.** $\left(-\frac{7}{8}\right) - \left(-\frac{1}{4}\right)$.

15. $\frac{11}{3} - \frac{47}{6}$.   **16.** $\frac{2}{5} - \frac{9}{7}$.

8.10   Absolute value

We shall now define an important concept called the **absolute value** of a rational number.

Definition 8.10.1. *Let f be any rational number. Then the **absolute value** of f, which we write as $|f|$, is defined as follows:*
   (1) $|f| = f$ *if $f \geq 0$;*
   (2) $|f| = -f$ *if $f < 0$.*
For example, $|7| = 7$, $|0| = 0$, $|-3| = -(-3) = 3$, and $\left|-\frac{2}{3}\right| = -\left(-\frac{2}{3}\right) = \frac{2}{3}$.

Notice that according to the definition, the absolute value of a number which is positive or zero is the number itself while the absolute value of a negative number is the negative of the negative number and hence is a positive number. Thus the absolute value of every rational number except zero is a positive number, and the absolute value of zero is zero.

On the number line, $|f|$ is the distance in units from the origin to the point f, considering the distance to be positive whether measured to the right or to the left. For example, the distance between the origin and the point $-3\frac{1}{2}$ is $\left|-3\frac{1}{2}\right| = 3\frac{1}{2}$ units.

By definition, $f > g$ if and only if there is positive number h such that $g + h = f$. But by definition of subtraction, $h = f - g$ and $g + (f - g) = f$. But $f = g + (f - g)$ is obtained by going $f - g$ units to the right of g on the number line. Thus, in this case, $f - g$ is the distance in units between the points f and g

on the number line. Similarly, when $f < g$, $g - f$ is positive and is the distance in units between f and g. When $f = g$, then f and g are the same point, and the distance between f and g is 0 units. In this latter case the distance in units between f and g is either $f - g$ or $g - f$ since these differences are both equal to zero.

In all cases, then, the distance in units between rational points f and g on the number line is $f - g$ or $g - f$. In fact when $f = g$, it is both of these and when $f \neq g$, it is the positive one of the two choices. Thus to get the distance in units between f and g on the number line, we subtract in the correct order.

Let us see what happens if we subtract in the wrong order. Take the points 3 and 7, for example. The distance between these points is $7 - 3 = 4$ units, but $3 - 7 = -4$, the negative of the correct number. Now, if we use the absolute value of the difference, we get the correct result independently of the order of subtraction; that is, $|7 - 3| = 4$ and also $|3 - 7| = |-4| = 4$.

This is always the case, because when $f \neq g$, $|f - g| = |g - f|$, and this is always the positive one of the two choices $f - g$ or $g - f$ and hence is the distance between f and g. When $f = g$, the distance is still $|f - g| = |g - f|$, because $|0| = 0$.

Example 1. The distance between the points $3\frac{1}{2}$ and $-2\frac{1}{2}$ on the number line is $|3\frac{1}{2} - (-2\frac{1}{2})| = |3\frac{1}{2} + 2\frac{1}{2}| = |6| = 6$ units, or it is $|-2\frac{1}{2} - 3\frac{1}{2}| = |-6| = 6$ units.

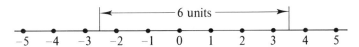

Example 2. The distance between the points $-5\frac{1}{2}$ and $-2\frac{1}{2}$ on the number line is $|-5\frac{1}{2} - (-2\frac{1}{2})| = |-5\frac{1}{2} + 2\frac{1}{2}| = |-3| = 3$ units, or it is $|-2\frac{1}{2} - (-5\frac{1}{2})| = |-2\frac{1}{2} + 5\frac{1}{2}| = |3| = 3$ units.

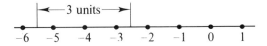

The idea of determining the absolute value of the difference to get a distance is often quite useful. The use of this and the number line helps greatly in " picturing " certain types of sets and in describing certain sets. For example, the set $\{x \in F \,|\, |x - 3| < 2\}$ is the set of all rational points x such that the distance between x and 3 is less than 2 units. One can draw a diagram of this set as follows:

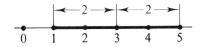

In this and other diagrams we shall understand that a solid dot • on a line means that the point is part of the set being pictured, while a small circle ○ means that the point is not a part of the set being pictured.

From the diagram we can see that this is the set of all rational points x between 1 and 5; that is, $\{x \in F \mid 1 < x < 5\}$.

The set $\{x \in F \mid |x - 3| > 2\}$ is the set of all rational points x whose distance from 3 is greater than 2 units. A diagram of this set is the following:

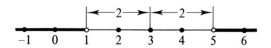

From the diagram we see that it is the set of all rational points that are either to the right of 5 or to the left of 1. This set can also be described as $\{x \in F \mid x < 1 \text{ or } x > 5\}$.

The set $\{x \in F \mid |x + 6| < 3\}$ is the set of all rational points x such that the distance between x and -6 is less than 3 units, because $x + 6 = x - (-6)$.

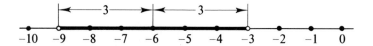

Thus this set can also be expressed in set notation as $\{x \in F \mid -9 < x < -3\}$. It is important to realize that it is the absolute value of a *difference*, not a sum, that is related to distance. In particular, $|x + 6|$ is the distance in units between x and -6, because $|x + 6| = |x - (-6)|$.

Since, on the number line, $|x| = |x - 0|$ is the distance in units between the origin and the point x, the values of x for which $|x| = 1$ are 1 and -1. That is, the points 1 and -1 are those and only those points whose distance from the origin is 1 unit.

Exercises

1. Find the absolute value of each of the following numbers:
   (a) $\frac{3}{2}$;   (b) $-\frac{7}{8}$;   (c) 0;   (d) -0;   (e) $\frac{2}{3} - \frac{7}{8}$.
2. If $u = v$, find (a) $|v - u|$;   (b) $|u - v|$.
3. If $u < v$, find (a) $|v - u|$;   (b) $|u - v|$.
4. If $u > v$, find (a) $|v - u|$;   (b) $|u - v|$.

For what values of x does each of the following hold?

5. $|x| = 6$;   6. $|x| = 0$;
7. $|-x| = 5$;   8. $|x| > 0$;
9. $|x| < -3$;   10. $|x - 2| = 0$;
11. $|x + 2| = 0$;   12. $|x - 3| < 0$.

For each of the following sets, first describe the set in words in terms of distance, then draw a diagram of the set on the number line if the set is not empty, and finally, express the set in set notation in another way.

13. $\{x \in F \mid |x| < 3\}$;

14. $\{x \in F \mid |x| > 2\}$;

15. $\{x \in F \mid |x - 2| < 1\}$;

16. $\{x \in F \mid |x - \frac{1}{2}| > 3\}$;

17. $\{x \in F \mid |x + 2| > 5\}$;

18. $\{x \in F \mid |x + \frac{2}{3}| < \frac{7}{8}\}$;

19. $\{x \in F \mid |x - 1| < 0\}$;

20. $\{x \in F \mid |x - 3| \leq 0\}$;

21. $\{x \in F \mid |x + 3| \leq 0\}$;

22. $\{x \in F \mid |x - 3| > -2\}$;

23. $\{x \in F \mid |x + 2| < -1\}$;

24. $\{x \in F \mid |x - 1| = 3\}$;

25. $\{x \in F \mid |x - 5| = 2\}$;

26. $\{x \in F \mid |x + 7| = 4\}$;

27. $\{x \in F \mid |x + 1| = -3\}$;

28. $\{x \in F \mid |x + 2| < 0\}$.

8.11   Solutions of equations and inequalities

Let us consider the set $S = \{x \in F \mid |x| = 3\}$. This set S can be described more simply by $S = \{3, -3\}$, because 3 and -3 are the only numbers whose absolute values are 3.

The set S is called the **solution set** of the equation $|x| = 3$ in F. When one finds the elements of this set, or at least a simple form for expressing the set, we say that he has **solved** the equation. The elements of the solution set are called the **solutions** of the equation. In this case the solutions are 3 and -3.

We now formalize the information given for the example $|x| = 3$. First we define the term solution of an equation or inequality.

Definition 8.11.1. *A number r of F is called a **solution** in F of an equation or inequality involving x if and only if it becomes a true statement when x is replaced by r.*

For example $\frac{2}{3}$ is a solution of $3x = 2$ because $3 \cdot \frac{2}{3} = 2$ is a true statement. Similarly -5 is a solution of $|x| = 5$ because $|-5| = 5$. The number 5 is also a solution of the equation $|x| = 5$. Also, 3 is a solution of the inequality $2x < 10$ because $2(3) < 10$, but there are many other solutions; in particular, 4, $\frac{2}{3}$, and -6 are others.

When a number r is a solution of an equation or inequality, one sometimes says that r *satisfies* it.

Definition 8.11.2. *The set of all solutions in F of an equation or inequality is called its **solution set** in F.*

Definition 8.11.3. *To **solve** an equation or inequality in F means to find its solution set in F.*

When one replaces the F in the three preceding definitions by W, I, or any other set of numbers, one obtains corresponding definitions for the solution set in W, I, or the other set. For example, the solution set of $3|x| = 2$ in W and in I is $\emptyset$, but is $\{\frac{2}{3}, -\frac{2}{3}\}$ in F. For the remainder of this section, all of the equations and inequalities to be solved are to be solved in F, and hence the "in F" part will be omitted.

Notice that according to Definitions 8.11.1 and 8.11.2 the solution set of $|x| = 3$ is $\{x \in F \mid |x| = 3\}$ and of $2x < 4$ is $\{x \in F \mid 2x < 4\}$. However, when we solve equations and inequalities, we will try to express them in their simplest form and hence will express these solution sets as $\{3, -3\}$ and $\{x \in F \mid x < 2\}$ respectively.

Examples:

(1) Solve the equation $|x| = 0$. The solution set is $\{0\}$.

(2) Solve the inequality $|x| < 3$. Here the solution set (which can be determined from the number line diagram) is $\{x \in F \mid -3 < x < 3\}$. In this case the solution set contains infinitely many elements; that is, the inequality has infinitely many solutions.

(3) Solve $|x| = -5$. Here the solution set is $\emptyset$. The equation has no solution because the absolute value of a rational number is always either positive or zero and is never negative.

(4) Solve $2(x + 1) = 2x + 2$. Here the solution set is F, because, for every rational number x, $2(x + 1) = 2x + 2$ by the distributive property. That is, every number of F is a solution of the equation. This is an example of an identity. When the solution set is the whole set under consideration, we often call the equation an **identity**.

Definition 8.11.4. *Two equations (or inequalities) which have the same solution set are called* ***equivalent***.

Examples:

(1) $x = 0$ and $|x| = 0$ are equivalent equations, both having the solution set $\{0\}$.

(2) $2x = 4$ and $x = 2$ are equivalent equations, both having $\{2\}$ as their solution set.

(3) $2x < 4$ and $x < 2$ are equivalent inequalities having the solution set $\{x \in F \mid x < 2\}$.

(4) $|x| = -5$ and $|x| = -2$ are equivalent with solution set $\emptyset$.

Now let us see how to go about solving certain equations and inequalities. One often solves an equation (or inequality) by changing it by successive steps to simpler and simpler equivalent equations (or inequalities) until the equation (or inequality) itself simply describes the solution set. By Theorems 8.4.2 and 8.4.4, there are four fundamental operations that can be performed on an equation to change it to an equivalent equation.

Fundamental Operations to Change an Equation to an Equivalent One.

1. *Add the same quantity to both sides of the equation.*
2. *Subtract the same quantity from both sides of the equation.*
3. *Multiply both sides of the equation by the same nonzero quantity.*
4. *Divide both sides of the equation by the same nonzero quantity.*

Similarly, by Theorem 8.6.3, there are four fundamental operations to change an inequality to an equivalent inequality.

Fundamental Operations to Change an Inequality to an Equivalent One.

1. *Add the same quantity to both sides of the inequality.*
2. *Subtract the same quantity from both sides of the inequality.*
3. *Multiply both sides of the inequality by the same nonzero quantity, but change the direction of the inequality if the nonzero quantity is negative.*
4. *Divide both sides of the inequality by the same nonzero quantity, but change the direction of the inequality if the nonzero quantity is negative.*

It is usually best to simplify until the left side of the equation (or inequality) is reduced to x, and the right side is as simple as possible.

Examples:

(1) Solve $4x + 3 = 7x - 8$.

First reduce this to a simpler equivalent equation by subtracting 3 from both sides to get

$$4x = 7x - 11.$$

Next subtract $7x$ from both sides to obtain the equivalent equation

$$-3x = -11.$$

Now divide both sides by -3, which is $\neq 0$, to get the equivalent equation

$$x = \frac{11}{3}.$$

We have reached the simplest form of the equation, and the solution set is $\{\frac{11}{3}\}$. Thus $x = \frac{11}{3}$ is the only solution.

(2) Solve $6x - 5 > 8x + 6$.

Add 5 to both sides to obtain the equivalent inequality

$$6x > 8x + 11.$$

Now subtract $8x$ from both sides to get

$$-2x > 11.$$

Next divide both sides by -2 (and reverse the direction of the inequality) to get

$$x < -\frac{11}{2}.$$

The solution set is $\{x \in F \mid x < -\frac{11}{2}\}$.

(3) Solve $2 - 3x < 5 + x < 4x - 1$.

This means $2 - 3x < 5 + x$ and $5 + x < 4x - 1$. These are equivalent to

$$-4x < 3 \quad \text{and} \quad -3x < -6.$$

So the given inequality is equivalent to

$$x > -\frac{3}{4} \quad \text{and} \quad x > 2.$$

But $x > -\frac{3}{4}$ and $x > 2$ are equivalent to just $x > 2$, so the solution set is $\{x \in F \mid x > 2\}$.

(4) Solve $\dfrac{1}{x - 2} = 0$.

To word this in another way, we are trying to find $\{x \in F \mid 1/(x - 2) = 0\}$. The statement $x - 2 \neq 0$ (i.e., $x \neq 2$) is actually included in the statement $1/(x - 2) = 0$ because denominators can never be zero; hence

$$\left\{x \in F \;\middle|\; \frac{1}{x - 2} = 0\right\} = \left\{x \in F \;\middle|\; x \neq 2 \text{ and } \frac{1}{x - 2} = 0\right\}.$$

Now when $x - 2 \neq 0$ (i.e., $x \neq 2$), $1/(x - 2) = 0$ is equivalent to (multiplying both sides by $x - 2$) $1 = 0$, so

$$\left\{x \in F \;\middle|\; \frac{1}{x - 2} = 0\right\} = \left\{x \in F \;\middle|\; x \neq 2 \text{ and } \frac{1}{x - 2} = 0\right\}$$

$$= \{x \in F \mid x \neq 2 \text{ and } 1 = 0\}.$$

This set is $\varnothing$ because $1 \neq 0$. Thus the solution set is $\varnothing$ and the given equation has no solution.

(5) Solve $2x + 4 < 5 + x < 4x - 1$.

This means $2x + 4 < 5 + x$ and $5 + x < 4x - 1$, which are equivalent to

$$x < 1 \quad \text{and} \quad -3x < -6,$$

which are equivalent to

$$x < 1 \quad \text{and} \quad x > 2.$$

So the given inequality is equivalent to $x < 1$ and $x > 2$. But $\{x \in F \mid x < 1$ and $x > 2\} = \varnothing$, so the solution set of the given inequality is $\varnothing$.

(6) Solve $2x = 3x$.

Subtract $2x$ from both sides to get the equivalent equation $0 = x$; that is, $x = 0$. The solution set is $\{0\}$.

(7) Solve $2x - 1 = 2x - 1$.

Here one does not have to change the equation at all to see that

$$\{x \in F \mid 2x - 1 = 2x - 1\} = F;$$

so every number in F is a solution, and the equation is an identity.

(8) Solve $3 = 3$ in F.

$\{x \in F \mid 3 = 3\} = F$; so every number in F is a solution, and we have another identity.

(9) Solve $|x - 2| < 3$.

This is most easily solved by the method of Section 8.10; that is,

$\{x \in F \mid |x - 2| < 3\} = \{x \in F \mid$ the distance between x and 2 is < 3 units$\}$. By looking at the number line shown in Figure 8.11.1,

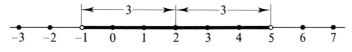

Figure 8.11.1

it is seen that this set is $\{x \in F \mid -1 < x < 5\}$.

Example (6) above points out that one must be especially careful when he multiplies or divides an equation by something containing x. This is because the quantity containing x may be zero for some x's and not equal to zero for others, and in order to get an equivalent equation, the quantity by which the members of an equation are multiplied or divided must not be zero. If we divide both sides of $2x = 3x$ by x we get $2 = 3$, an equation whose solution set is $\varnothing$, whereas the original equation has solution set $\{0\}$.

Exercises

Solve the following in F:

1. $5x + 7 = 8x - 2$.

2. $\frac{3}{2}x - \frac{7}{3} = \frac{5}{8}x + \frac{2}{3}$.

3. $5x + 7 < 8x - 2$.

4. $\frac{3}{2}x - \frac{7}{3} > \frac{5}{6}x + \frac{2}{3}$.

5. $3x - 7 < 4x + 8 < 2x - 3$.

6. $3 - 8x < 2x + 5 < 1 - x$.

7. $8 + 3x > 2x - 11 > 4 - x$.

8. $\dfrac{1}{x + 1} = 0$.

9. $-3x < 12$.

10. $|x - 5| < 6$.

11. $|x| < 3$.

12. $|x| > 2$.

13. $|x + 6| < 3$.

14. $|x + 4| > 3$.

15. $|2 - x| < 5$.

16. $7x = 2x$.

17. $3 = 5$.

18. $6 = 6$.

19. $2x = 2x$.

20. $|-x| < 5$.

21. $|x - 7| = 3$.

22. $|x + 6| = 2$.

chapter 9

The real numbers

Before defining real numbers we shall consider some preliminary material on exponents and decimals.

9.1 Exponents

In Section 4.6 we defined a^h for all whole numbers a and h, except when a and h are both zero. When $a = h = 0$, no worthwhile definition can be given. We will now extend the definition to define a^h when a is any rational number and h is any integer, except for the cases where $a = 0$ and the exponent is not positive.

Definition 9.1.1. *If a is a rational number and h is an integer, then*
    (1) *If $h > 0$, $a^h = $ the product of h factors each equal to a;*
    (2) *If $h = 0$ and $a \neq 0$, $a^h = 1$;*
    (3) *If $-h < 0$ and $a \neq 0$, $a^{-h} = \dfrac{1}{a^h}$.*

For example:

    (1) $\left(\dfrac{2}{3}\right)^3 = \dfrac{2}{3} \cdot \dfrac{2}{3} \cdot \dfrac{2}{3} = \dfrac{8}{27}.$ (2) $\left(-\dfrac{1}{3}\right)^2 = \left(-\dfrac{1}{3}\right)\left(-\dfrac{1}{3}\right) = \dfrac{1}{9}.$

(3) $\left(-\dfrac{2}{3}\right)^{3} = \left(-\dfrac{2}{3}\right)\left(-\dfrac{2}{3}\right)\left(-\dfrac{2}{3}\right) = -\dfrac{8}{27}$.

(4) $\left(-\dfrac{7}{3}\right)^{0} = 1$.   (5) $2^{-3} = \dfrac{1}{2\cdot 2\cdot 2} \overset{\bullet}{=} \dfrac{1}{8}$.

(6) $\left(\dfrac{3}{2}\right)^{-2} = \dfrac{1}{\left(\dfrac{3}{2}\right)\left(\dfrac{3}{2}\right)} = \dfrac{1}{\dfrac{9}{4}} = 1\cdot\dfrac{4}{9} = \dfrac{4}{9}$.

(7) $10^{-3} = \dfrac{1}{10^{3}} = \dfrac{1}{1000}$.   (8) $\dfrac{1}{a^{-h}} = \dfrac{1}{\dfrac{1}{a^{h}}} = 1\cdot\dfrac{a^{h}}{1} = a^{h}$.

Notice that if $a = 0$ and h is positive, we cannot define $a^{-h} = 1/a^{h}$ because it would then be $1/0$ which has no meaning. There is, in fact, no useful way to define a^{-h} when $a = 0$ and h is a positive whole number or zero, so we leave these cases undefined.

The three properties of Theorem 4.6.1, hold for exponents defined by this extended definition. We include a fourth property also.

Theorem 9.1.1. *If a, b ∈ F and r and s are integers, and if both sides of the equation have a meaning, then*

(1) $a^{r} \cdot a^{s} = a^{r+s}$;

(2) $(a^{r})^{s} = a^{rs}$;

(3) $(ab)^{r} = a^{r} b^{r}$;

(4) $\dfrac{a^{r}}{a^{s}} = a^{r-s}$.

We will not give a complete proof of this theorem, but will illustrate the proof with several examples.

Examples:

(1) $3^{-2} \cdot 3^{5} = \dfrac{1}{3\cdot 3}\cdot 3\cdot 3\cdot 3\cdot 3\cdot 3 = \dfrac{3\cdot 3\cdot 3\cdot 3\cdot 3}{3\cdot 3} = 3^{3} = 3^{-2+5}$.

(2) $5^{-2} \cdot 5^{-3} = \dfrac{1}{5\cdot 5}\cdot\dfrac{1}{5\cdot 5\cdot 5} = \dfrac{1}{5^{5}} = 5^{-5} = 5^{-2+(-3)}$.

(3) $(6^{5})^{-3} = \dfrac{1}{6^{5}\cdot 6^{5}\cdot 6^{5}} = \dfrac{1}{6^{15}} = 6^{-15} = 6^{(5)(-3)}$.

(4) $\dfrac{7^{5}}{7^{2}} = \dfrac{7\cdot 7\cdot 7\cdot 7\cdot 7}{7\cdot 7} = 7^{3} = 7^{5-2}$.

(5) $\dfrac{7^3}{7^{-2}} = 7 \cdot 7 \cdot 7 \div \dfrac{1}{7 \cdot 7} = 7 \cdot 7 \cdot 7 \cdot 7 \cdot 7 = 7^5 = 7^{3-(-2)}$.

Note that if a and b are nonzero integers and h is a positive integer, then

$$\left(\frac{a}{b}\right)^{-h} = \frac{1}{\left(\dfrac{a}{b}\right)^h} = \frac{1}{\dfrac{a^h}{b^h}} = 1 \cdot \frac{b^h}{a^h} = \frac{b^h}{a^h} = \left(\frac{b}{a}\right)^h.$$

That is,

$$\left(\frac{a}{b}\right)^{-h} = \left(\frac{b}{a}\right)^h$$

For example, $\left(\dfrac{5}{6}\right)^{-2} = \left(\dfrac{6}{5}\right)^2$ and $\left(\dfrac{-2}{7}\right)^{-4} = \left(\dfrac{7}{-2}\right)^4 = \left(\dfrac{7}{2}\right)^4$,

while $\left(\dfrac{-3}{5}\right)^{-3} = \left(\dfrac{5}{-3}\right)^3 = \left(-\dfrac{5}{3}\right)^3 = -\left(\dfrac{5}{3}\right)^3$.

Exercises

Simplify the following:

1. $\left(-\dfrac{3}{4}\right)^2$;   2. $\left(-\dfrac{4}{5}\right)^{-2}$;   3. $\left(\dfrac{3}{7}\right)^0$;   4. $\dfrac{1}{5^{-2}}$;

5. $\left(\dfrac{3}{2}\right)^{-4}$;   6. $(-2)^{-4}(-2)^7$;   7. $\left(\dfrac{3}{4}\right)^{-3}\left(\dfrac{4}{3}\right)^2$;   8. $(5^3)^{-2}$;

9. $\dfrac{3^4}{3^6}$;   10. $\dfrac{2^{-4}}{2^{-7}}$;   11. $(2^{-3} \cdot 3^{-2})^2$;   12. $\dfrac{(\frac{2}{3})^{-2}}{(\frac{3}{2})^4}$;

13. $\dfrac{(2^{-4})^2}{(2^3)^4}$;   14. $\left[\left(-\dfrac{2}{3}\right)^{-5} \times \left(\dfrac{3}{2}\right)^{-3}\right] \div \left(\dfrac{2}{3}\right)^2$;   15. $2(10^{-5} \times 10^{-2})$;

16. $[2(10)^2 + 3(10) + 5 + 7(10)^{-1} + 9(10)^{-2}] \times 10^4$;
17. $[3(10) + 2 + (10)^{-1}] \times 10^{-3}$;
18. $[3(10)^3 + 0(10)^2 + 0(10) + 5 + 6(10)^{-1}] \times 10^{-2}$;

19. $\left[\left(-\dfrac{5}{3}\right)^{-2}\right]^{-1}$;   20. $a^{-3} \cdot a^5 \cdot a^4 \cdot a^{-7}$;   21. $(a^{-3}b^4)^5$;

22. $(a^3 b^2)^{-3}(a^2 b^{-2})^4$;   23. $10^{-3} \times 10^3$;   24. $10^7 \times 10^{-7}$;
25. $6^8 \times 6^{-8}$;   26. $(10^{-4})^{-1}$;   27. $(a^{-2}b^{-3})^{-4}$.

9.2 Decimal notation for rational numbers

Toward the end of Section 5.1, we considered the Hindu-Arabic numeral system of expressing a whole number as a sum of terms each of which is a power of 10 times one of the numbers 0, 1, 2, ..., 9; that is, in the form

$$a_k(10^k) + a_{k-1}(10)^{k-1} + \cdots + a_2(10)^2 + a_1(10) + a_0$$

where each of the a's is 0, 1, 2, 3, 4, 5, 6, 7, 8, or 9. We then expressed the number as

$$a_k\, a_{k-1} \cdots a_2\, a_1\, a_0.$$

We will now extend this system to include negative powers of 10 and negative numbers. We shall consider numbers of the form

$$\pm[a_k(10)^k + a_{k-1}(10)^{k-1} + \cdots + a_2(10)^2 + a_1(10)$$
$$+ a_0 + a_{-1}(10)^{-1} + a_{-2}(10)^{-2} + \cdots],$$

where k is an integer and each of the a's is one of the numbers 0, 1, 2, $\cdots$, 9.
We shall understand

$$+[a_k(10)^k + a_{k-1}(10)^{k-1} + \cdots + a_1(10) + a_0$$
$$+ a_{-1}(10)^{-1} + a_{-2}(10)^{-2} + \cdots]$$

to mean

$$a_k(10)^k + a_{k-1}(10)^{k-1} + \cdots + a_1(10) + a_0 + a_{-1}(10)^{-1} + a_2(10)^{-2} + \cdots$$

and

$$-[a_k(10)^k + a_{k-1}(10)^{k-1} + \cdots + a_1(10) + a_0$$
$$+ a_{-1}(10)^{-1} + a_{-2}(10)^{-2} + \cdots]$$

to mean

$$-a_k(10)^k - a_{k-1}(10)^{k-1} - \cdots - a_1(10) - a_0 - a_{-1}(10)^{-1} - a_{-2}(10)^{-2} - \cdots.$$

We shall write a number of this form as

$$\pm a_k\, a_{k-1} \cdots a_2\, a_1\, a_0.a_{-1}\, a_{-2} \ldots,$$

marking the space between a_0 and a_{-1} with a dot called the **decimal point**.

For example, 278.305 means

$$2(10)^2 + 7(10) + 8 + 3(10)^{-1} + 0(10)^{-2} + 5(10)^{-3},$$

and -278.305 means

$$-[2(10)^2 + 7(10) + 8 + 3(10)^{-1} + 0(10)^{-2} + 5(10)^{-3}].$$

Numerals of this form are called **decimals**. The decimal point is used to indicate where the integer part ends. The numbers 0, 1, 2, ..., 9 occurring in a decimal are called the **digits**.

It is important to notice the effect on the position of the decimal point when a number is multiplied by a power of 10. For example:

(1) $32.681 \times 10^2 = [3(10)^1 + 2 + 6(10)^{-1} + 8(10)^{-2} + 1(10)^{-3}] \times 10^2$
$= 3(10)^3 + 2(10)^2 + 6(10) + 8 + 1(10)^{-1} = 3268.1$

(2) $32.681 \times 10^{-2} = [3(10)^1 + 2 + 6(10)^{-1} + 8(10)^{-2} + 1(10)^{-3}] \times 10^{-2}$
$= 3(10)^{-1} + 2(10)^{-2} + 6(10)^{-3} + 8(10)^{-4} + 1(10)^{-5} = .32681$

(3) $-32.681 \times 10^3 = -[3(10)^1 + 2 + 6(10)^{-1} + 8(10)^{-2} + 1(10)^{-3}]$
$\times 10^3 = -[3(10)^4 + 2(10)^3 + 6(10)^2 + 8(10) + 1] = -32681.$

It is seen from the first of these examples that when a number is multiplied by 10^2, each of the exponents of 10 is increased by 2 and, hence, the decimal point is moved 2 places to the right. Similarly, multiplying by 10^3 increases each exponent by 3, and, hence, moves the decimal point 3 places to the right. In general, *multiplying by 10^s for any positive whole number s moves the decimal point s places to the right.* From the second example, we see that when a number is multiplied by 10^{-2} the decimal point is moved 2 places to the left, and, in general, *multiplying by 10^{-s} for any positive whole number s moves the decimal point s places to the left.*

As further illustrations, notice that

(1) $6.82 \times 10^4 = 68,200$ (2) $3.71824 \times 10^3 = 3718.24$
(3) $0.00247 \times 10^4 = 24.7$ (4) $-6.82 \times 10^{-3} = -0.00682$
(5) $397.62 \times 10^{-3} = 0.39764$ (6) $0.0178 \times 10^{-3} = 0.0000178$
(7) $-0.0178 \times 10^6 = -17,800$ (8) $63.471 = 63471 \times 10^{-3}$
(9) $861.792 = 861792 \times 10^{-3}$ (10) $0.31007 = 31007 \times 10^{-5}.$

Notice in particular that every decimal can be expressed as a decimal with just one digit to the left of the decimal point times a power of 10. For example,

(1) $379.26 = 3.7926 \times 10^2,$ (2) $0.00723 = 7.23 \times 10^{-3},$
(3) $-0.000000526 = -5.26 \times 10^{-7},$
(4) $0 = 0.0,$ (5) $3 = 3.0 \times 10^0,$
(6) $-8427000 = -8.427 \times 10^6,$ (7) $79200000000000 = 7.92 \times 10^{13}.$

This way of writing a decimal is often called *scientific notation* and is useful in many instances because it requires less space and because it makes counting 0's unnecessary.

Now let us see how to express a rational number as a decimal. The following examples illustrate how it can be done.

(1) $\dfrac{1}{2} = \dfrac{1}{2} \times 10 \times 10^{-1} = \dfrac{10}{2} \times 10^{-1} = 5 \times 10^{-1} = .5,$

(2) $\dfrac{1}{4} = \dfrac{1}{4} \times 10^2 \times 10^{-2} = \dfrac{100}{4} \times 10^{-2} = 25 \times 10^{-2} = .25,$

(3) $\dfrac{1}{10} = 10^{-1} = 1 \times 10^{-1} = .1,$

(4) $\dfrac{1}{100} = 10^{-2} = 1 \times 10^{-2} = .01,$

(5) $\dfrac{21}{8} = \dfrac{21}{8} \times 10^3 \times 10^{-3} = \dfrac{21000}{8} \times 10^{-3} = 2625 \times 10^{-3} = 2.625,$

(6) $\dfrac{63}{25} = \dfrac{63}{25} \times 10^2 \times 10^{-2} = \dfrac{6300}{25} \times 10^{-2} = 252 \times 10^{-2} = 2.52.$

The technique is to multiply the fraction by the proper $10^k \times 10^{-k} = 1$. The 10^k puts k zeros on the end of the numeral in the numerator; that is, it moves the decimal point k places to the right. Division is performed and then the 10^{-k} moves the decimal point back to where it started, k places to the left. It is not necessary to know the proper power of 10 in advance; one zero at a time is attached at the end of the numeration until the division comes out evenly. As an illustration, we shall do this for 63/25 one step at a time as follows, to obtain the result $252 \times 10^{-2} = 2.52$.

```
                2                            25                            252
Step 1.   25)63             Step 2.   25)630            Step 3.   25)6300
               50                           50                            50
               13                          130                           130
                                           125                           125
                                             5                            50
                                                                          50
```

Since the decimal point is positioned k places to the left after k zeros have been attached to the numerator, one does not have to count to get its position, and the calculation could appear as follows:

```
              2.52
      25)63.00
         50
         13 0
         12 5
            50
            50
```

The calculation indicates that

$$\dfrac{63.00}{25} = 2.52.$$

The decimal point in the quotient always occurs directly above the decimal point in the dividend.

However, it is possible (and in fact likely) that the division does not come out evenly no matter how many zeros are attached to the numerator. Consider $\frac{1}{3}$, for example. The first two steps are

$$
\begin{array}{r}
.33 \\
3\overline{)1.00} \\
9 \\
\hline
10 \\
9 \\
\hline
1
\end{array}
$$

Each time we add a new zero, we divide, getting another 3 and another remainder of 1. As the steps continue we get

$$
\begin{array}{r}
.3333\ldots \\
3\overline{)1.0000\ldots}
\end{array}
$$

and hence

$$
\frac{1}{3} = 0.3333\ldots.
$$

The 3's continue without stopping.

As another example, consider $\frac{4}{330}$. As we divide, we get

$$
\begin{array}{r}
.01212 \\
330\overline{)4.00000} \\
3\ 30 \\
\hline
700 \\
660 \\
\hline
400 \\
330 \\
\hline
700
\end{array}
$$

One sees that the decimal continues, with 12 repeating over and over again. We have

$$
\frac{4}{330} = 0.012121212\ldots.
$$

We will write the decimal with the 12 repeating as $0.0\overline{12}$, the bar over the 12 indicating that the 12 forever repeats. Thus

$$
\frac{4}{330} = 0.0\overline{12}.
$$

Similarly

$$\frac{334}{330} = 1.0\overline{12} \quad \text{and} \quad \frac{246}{700} = 0.35\overline{142857}$$

Now that we have considered some particular decimals we will find it convenient to name certain types of them.

Definition 9.2.1. *A decimal is called:*

(1) *a **terminating decimal**, if there is a digit to the right of which there are no nonzero digits;*

(2) *a **nonterminating decimal** if it is not a terminating decimal;*

(3) *a **repeating decimal** if there is a digit to the right of which the same sequence of a certain number of digits is repeated over and over again without ever stopping;*

(4) *a **nonrepeating decimal** if it is not a repeating decimal.*

Examples:

(1) terminating decimals: 62.471, 7, 0, -86, -3.07171, 0.0025;

(2) nonterminating decimals: $-3.0\overline{71}$, $6.\overline{8}$, $27.3\overline{168}$, $0.0\overline{2}$, in which a line or "bar" above a sequence of digits indicates that that sequence is repeated without end;

(3) repeating decimals: $6.0\overline{0} = 6$, $-7 = -7.\overline{0}$, $0.0\overline{2}$, $-3.0\overline{71}$;

(4) nonrepeating decimals: 3.12112111211112... where the 2's are next separated by five 1's, then by six, then by seven, etc.,...; 86.0100110001110000-1111... where the pattern continues.

According to these definitions, every terminating decimal is a repeating decimal because the zeros repeat from some point on. Also every nonrepeating decimal is a nonterminating decimal.

The number of **decimal places** and the number of **places** of a terminating decimal are important considerations. These terms are defined as follows:

Definition 9.2.2. *If t is a terminating decimal, then*

(1) *the number of **decimal places** of t is the number of digits written to the right of the decimal point;*

(2) *the number of **places** of t is the number of digits from the first nonzero digit to the last written digit.*

For example 34.26 is a 2 decimal place numeral and a 4 place numeral, while 0.00270 is a 5 decimal place numeral and a 3 place numeral. Also 0.0027 is a 4 decimal place numeral and a 2 place numeral.

We have seen several examples of rational numbers whose decimal representation is either a terminating decimal or a repeating decimal. This is in fact true of all rational numbers as the following theorem indicates.

Theorem 9.2.1. *The decimal representation of a rational number is a repeating decimal (either terminating or nonterminating).*

Proof. We will prove that if the decimal representation of a rational number is not a terminating decimal, then it is a repeating decimal.

Let a/d, where a and d are integers and $d > 0$, be any rational number in lowest terms. In the division process of determining the decimal representation of a/d, consider the remainders obtained as the division is carried out. Each remainder is greater than or equal to zero and less than d; therefore, there are only d possible remainders. After we get to the right of the decimal point in the division, each step is carried out by dividing the last remainder, with one or more zeros attached, by d. Since there are at most d different remainders possible, any remainder that has been obtained will occur again after at most k steps, for some whole number $k \leq d$. As the process continues from this point the remainders must follow the same pattern in the next k steps and then the next k steps, etc. Hence, the same succession of k digits must repeat over and over again without end.

As an illustration of the proof just given, let us consider the decimal representation of $\frac{31}{168}$. The first few steps in the division are as follows:

$$
\begin{array}{r}
.184\ldots \\
168\overline{)31.000\ldots} \\
16\ 8 \\
\hline
14\ 20 \\
13\ 44 \\
\hline
760 \\
672 \\
\hline
88
\end{array}
$$

The first 8 remainders (only the first 3 of which appear above) are 142, 76, 88, 40, 64, 136, 16, 160. Like all remainders when division by 168 is performed, they are greater than or equal to 0 and less than 168, and we see that there are at most 168 distinct remainders possible. If a remainder in the division process were zero then we should have a terminating decimal representation. If no remainders were zero, then after at most 167 steps, we would get a remainder we had obtained before; the whole sequence of remainders would then be repeated over and over again. In our specific example the 9th remainder is 88 and hence the next one must be 40, then 64, 136, 16, 160, 88, 40, 136, 16, etc. Thus the digits obtained from these remainders must repeat over and over again. In fact,

$$
\frac{31}{168} = 0.184\overline{523809}
$$

To show this in detail is left as one of the exercises.

The converse of Theorem 9.2.1 also holds, as we shall prove in Section 9.6. The converse states that *every repeating decimal equals a rational number.*

Exercises

Express each of the following rational numbers as a decimal. Use the "bar notation" if the decimal repeats from some point on.

1. $\dfrac{2}{3}$;

2. $\dfrac{1}{6}$;

3. $\dfrac{1}{8}$;

4. $\dfrac{1}{40}$;

5. $\dfrac{17}{11}$;

6. $\dfrac{12}{7}$;

7. $\dfrac{43}{130}$;

8. $\dfrac{1}{100}$;

9. $\dfrac{1}{200}$;

10. $\dfrac{1}{1000}$;

11. $\dfrac{31}{168}$;

12. $\dfrac{1}{300}$.

Express each of the following numbers in scientific notation.

13. 671.3;

14. 0.0849;

15. 0.0000065;

16. 3,700,000;

17. .07;

18. 68,010.

9.3   Calculations with terminating decimals

Many calculations are made using terminating decimals and we shall now consider addition, subtraction, multiplication, and division of terminating decimals.

Notice that every terminating decimal can be expressed as an integer times a power of 10 and hence is a rational number. For example

$$(1)\quad 64.8271 = 648271 \times 10^{-4} = \frac{648271}{10000}.$$

$$(2)\quad -.038 = -38 \times 10^{-3} = -\frac{38}{1000}.$$

Using this property we can add, subtract, multiply, or divide terminating decimals by first performing the corresponding operation on integers. To add, we also use the distributive property. For example:

$$64.8271 + 0.038 = 648271 \times 10^{-4} + 380 \times 10^{-4}$$
$$= (648271 + 380)10^{-4} = 648651 \times 10^{-4} = 64.8651.$$

In order to be able to use the distributive property, we expressed each decimal as an integer times the *same* power of 10. Then we added 648271 and 380 and used the 10^{-4} to move the decimal point 4 places to the left. This process can be written more easily in the following familiar way, where the decimal point in the second number is placed directly below the decimal point in the first number:

$$64.8271$$
$$.0380$$
$$\overline{64.8651}$$

Subtraction is very similar to addition. For example:

$$8.76 - 3.097 = 8760 \times 10^{-3} - 3097 \times 10^{-3}$$
$$= (8760 - 3097)10^{-3} = 5663 \times 10^{-3} = 5.663.$$

The calculation is usually written in the familiar form, where again, we are careful to place the decimal points in the same vertical column.

$$8.760$$
$$-3.097$$
$$\overline{5.663}$$

Similarly,

$$41.32 - 59.476 = -(59.476 - 41.82) = -17.656.$$

Multiplication is illustrated as follows:

$$38.71 \times 0.052 = 3871 \times 10^{-2} \times 52 \times 10^{-3} = 3871 \times 52 \times 10^{-2} \times 10^{-3}$$
$$= 3871 \times 52 \times 10^{-5} = 201292 \times 10^{-5} = 2.01292.$$

Here the term 3871×52 is obtained by removing the decimal points in 38.71×0.052, and we merely multiply 3871×52 and then use the 10^{-5} to place the decimal point. The -5 is obtained by adding the -2 and the -3. That is, the number of decimal places, 5, in the product 38.71×0.052 is the sum, $2 + 3$, of the numbers of decimal places in the factors. This is always the case, for if a and b are any integers and r and s are any whole numbers, then

$$a \times 10^{-r} \times b \times 10^{-s} = ab \times 10^{-(r+s)}.$$

Division of terminating decimals can also be done by first dividing in the system of integers and then putting the decimal point in the proper place. In fact, a quotient with terminating decimals in the numerator and denominator is equal to a quotient of two integers. As an illustration consider $68.51 \div 4.3$.

$$68.51 \div 4.3 = \frac{68.51}{4.3} = \frac{68.51 \times 10^2}{4.3 \times 10^2} = \frac{6851}{430}.$$

We merely have to multiply the numerator and denominator by a large enough power of 10 to move the decimal point to the end of the longest decimal. When doing this, one should be careful to put the necessary number of zeros on the other decimal if any are needed.

The first few steps of the division might look as follows:

$$
\begin{array}{r}
15.9 \\
430\overline{)6851.0} \\
430 \\
\hline
2551 \\
2150 \\
\hline
401\ 0 \\
387\ 0 \\
\hline
14\ 0
\end{array}
$$

or, to avoid using extra zeros, the first few steps could be written more simply as

$$
\begin{array}{r}
15.9 \\
43\overline{)685.1} \\
43 \\
\hline
255 \\
215 \\
\hline
40\ 1 \\
38\ 7 \\
\hline
1\ 4
\end{array}
$$

which indicates that we are replacing

$$
\frac{68.51}{4.3} \quad \text{by} \quad \frac{68.51 \times 10}{4.3 \ \times 10} = \frac{685.1}{43}.
$$

In general, in the simplified form, the decimal point of the quotient always appears k places to the right of the decimal point of the numerator, where k is the number of decimal places of the denominator.

We have now seen how to add, subtract, multiply, and divide terminating decimals by expressing them as integers times suitable powers of 10 and then performing the operation in the system of integers and finally placing the decimal point in the proper place.

The following theorem indicates the closure properties of the terminating decimals under the four fundamental operations.

Theorem 9.3.1. *The system of terminating decimals is closed under addition, subtraction, and multiplication, but is not closed under division.*

Proof. The sum or difference of two terminating decimals can be obtained by expressing both of them as an integer times the same power of 10. The distributive property is then used to express the sum or difference as the sum or difference of integers times a power of 10. Since the system of integers is closed under addition and subtraction, we then have an integer times a power of 10 which is a terminating decimal.

The product of two terminating decimals can also be obtained by expressing each of them as an integer times a power of 10 and then multiplying the integers obtained and the powers of 10 obtained to get an integer times a power of 10 which, again, is a terminating decimal.

To show that the terminating decimals are not closed under division, we only have to show that there exist two nonzero terminating decimals whose quotient is not a terminating decimal. Take for example $0.1 \div 0.3 = 1 \div 3$ $= \frac{1}{3} = 0.333\overline{3}...$, a nonterminating decimal.

Exercises

Find the terminating decimal that is equal to each of the following:

1. $67.281 + 1.9864$;   **2.** $67.281 - 1.9864$;

3. $1.9864 - 67.281$;   **4.** $-0.037 + 0.00937$;

5. 3.69×0.00972;   **6.** 87.42×192.6;

7. 34798.211×0.00021;   **8.** $(-0.371)(6.83)$.

Find the following quotients, expressing your answer as a terminating decimal or as a nonterminating repeating decimal.

9. $\dfrac{0.08}{0.00005}$;   **10.** $-\dfrac{0.00001}{0.002}$;   **11.** $\dfrac{36.8}{0.0003}$;

12. $\dfrac{0.000274}{1.3}$;   **13.** $\dfrac{0.0052}{0.000021}$;   **14.** $\dfrac{3.76}{2.5}$.

Change each of the following percents to a decimal and to a fraction (The term *percent*, often written %, means "hundredths" or "per hundred"; for example, $10\% = 10/100 = 10 \times 10^{-2} = .10 = 1/10$, and $6\frac{1}{2}\% = 6\frac{1}{2}/100 = 6.5/100 = 6.5 \times 10^{-2} = 0.065$):

15. 20%;   **16.** $33\frac{1}{3}\%$;   **17.** 5%;   **18.** 1%;   **19.** 0.01%;

20. $1/2\%$;   **21.** 0.002%;   **22.** 100%;   **23.** 150%;   **24.** $x\%$.

Change each of the following fractions to a percent:

25. $1/2$;   **26.** $1/4$;   **27.** $2/3$;   **28.** $2\frac{1}{2}$;   **29.** $1/100$.

Change each of the following decimals to a percent:

30. 0.3;   **31.** 0.007;   **32.** 3.6;   **33.** 0.64;   **34.** 2.05.

9.4   Approximations and rounding off decimals

When we express $\frac{2}{3}$ as a decimal we obtain

$$\frac{2}{3} = 0.666... = 0.\overline{6}$$

Notice that

$$0.6 = \frac{6}{10} = \frac{3}{5} \neq \frac{2}{3} \quad \text{but} \quad \frac{2}{3} - 0.6 = \frac{2}{3} - \frac{3}{5} = \frac{1}{15} < \frac{1}{10}.$$

Thus although $\frac{2}{3} \neq 0.6$, 0.6 differs from $\frac{2}{3}$ by a small number, in fact a number less than $\frac{1}{10} = 10^{-1}$. The number 0.6 is a **one-decimal approximation** to $\frac{2}{3}$. The number 0.66 is a **two-decimal approximation** to $\frac{2}{3}$ and

$$\frac{2}{3} - 0.66 = \frac{2}{3} - \frac{66}{100} = \frac{2}{300} < \frac{1}{100} = 10^{-2},$$

so 0.66 differs from $\frac{2}{3}$ by an amount less than $\frac{1}{100}$. It is a better approximation to $\frac{2}{3}$ than 0.6 is. Similarly 0.666 is a three-decimal approximation to $\frac{2}{3}$ and

$$\frac{2}{3} - 0.666 = \frac{2}{3} - \frac{666}{1000} = \frac{2}{3000} < \frac{1}{1000} = 10^{-3}$$

The farther we go in the decimal expansion of $\frac{2}{3}$ the better our approximation; in fact, if we take the first n decimal places, we get an approximation that differs from $\frac{2}{3}$ by an amount less than 10^{-n}.

This same situation occurs for any decimal. The terminating decimal obtained by taking only the first n decimal places of a decimal differs from it by an amount less than 10^{-n}. Hence the more decimal places taken, the better the approximation. In fact, we can choose a terminating decimal as close to a given nonterminating decimal as we wish by going out far enough in its decimal representation. For example we can choose a terminating decimal that differs from $\frac{2}{3} = 0.666... = 0.\overline{6}$ by a number less than $1/1,000,000 = 10^{-6}$ by taking the first 6 decimal places of its decimal representation; that is, 0.666666.

This is not the only terminating decimal that differs from $\frac{2}{3}$ by a number less than 10^{-6} however; for example, 0.666666201 and 0.666667 are such terminating decimals. While 0.6 is *a* one-decimal approximation to $\frac{2}{3}$, and

0.66 is *a* two-decimal approximation, they are not the best such approximations. The number 0.7 is the closest one-decimal approximation to $\frac{2}{3}$. It
differs from $\frac{2}{3}$ by a number smaller than 0.6 does. In fact,

$$0.7 - \frac{2}{3} = \frac{7}{10} - \frac{2}{3} = \frac{1}{30} \quad \text{and} \quad \frac{2}{3} - 0.6 = \frac{1}{15}.$$

The two-decimal place number closest to $\frac{2}{3}$ is 0.67. We say that 0.7 is *the* one-
decimal approximation to $\frac{2}{3}$ and 0.67 is *the* two-decimal approximation to $\frac{2}{3}$.
These are examples of **rounding off** decimals.

Definition 9.4.1. *To **round off** a given decimal to k decimal places means to
find the number with k decimal places that is closest to the given number. The
resulting k decimal place number is called **the k-decimal approximation** to the
given number.*

For example let us round off the number 63.175283 to various numbers of
decimal places.

    (1) The one-decimal approximation is 63.2.
    (2) The two-decimal approximation is 63.18.
    (3) The three-decimal approximation is 63.175.
    (4) The four-decimal approximation is 63.1753.
    (5) The five-decimal approximation is 63.17528.

What 2-decimal-place number is nearest to 2.725? Actually in this case, two
of them are equally near, namely 2.72 and 2.73. This situation occurs only
when rounding off to one less decimal place, a terminating decimal whose last
decimal place is 5. For uniformity, we will agree to take the even digit in such
a situation.

Example 1. 61.385 rounded to 2 decimal places is 61.38.

Example 2. 2.795 rounded to 2 decimal places is 2.80.

We use approximations to numbers frequently in our daily living. Most
measuring devices do not measure exact amounts but merely obtain approximations which are close enough to the exact amounts to make them useful.
When we measure a room, weigh ourselves, add 2 cups of sugar, or calculate
automobile gas mileage, we use approximations.

If we have approximations to two numbers, say r and s, then we can get
approximations to $r + s$, $r - s$, rs, and $r \div s$.

For addition and subtraction, the number of **decimal places** of the
approximations of r and s is important to the accuracy of resulting approximations to $r + s$ and $r - s$, while for multiplication and division, the number
of decimal places of r and s is not so important. The important thing for an
approximation to rs or $r \div s$ is the number of **places** of the approximations
to r and s.

In general, if we add or subtract reasonably good n-decimal-place approximations to r and s, we get a reasonably good n-decimal-place approximation to $r + s$ or $r - s$. This is because the decimal places after the nth place have little if any effect on the nth decimal place of $r + s$ or $r - s$.

For example, let $r = \frac{5}{3} = 1.6 = 1.6666\ldots$ and $s = \frac{1}{11} = 0.\overline{09} = 0.0909\ldots$. Then $r + s = \frac{58}{33} = 1.\overline{75} = 1.7575\ldots$. If we add 3-decimal place approximations to r and s, we get

$$1.667 + 0.091 = 1.758,$$

which is a good 3-decimal-place approximation to $1.7575\ldots$. Adding 4-decimal-place approximations to r and s, we obtain

$$1.667 + 0.0909 = 1.7576,$$

a good 4-decimal-place approximation to $r + s$. Similarly we have

$$1.667 - 0.091 = 1.576$$

and

$$1.6667 - 0.0909 = 1.5758,$$

which are good 3- and 4-decimal-place approximations to $\frac{5}{3} - \frac{1}{11} = \frac{52}{33} = 1.\overline{57} = 1.575757\ldots$.

When we multiply and divide reasonably good n-place approximations to r and s, and round off the results to n-places, we get reasonably good n-place approximations to rs and $r \div s$. For example: Let $r = \frac{138}{11} = 12.\overline{54} = 12.5454\ldots$, and $s = \frac{10}{9} = 1.\overline{1} = 1.111\ldots$. From these we calculate 1-, 2-, 3-, 4-, and 5-place approximations to rs, to obtain (we will use the symbol $\approx$ to mean "is approximately equal to"):

$$10 \times 1 = 10,$$
$$13 \times 1.1 = 14.3 \approx 14,$$
$$12.5 \times 1.11 = 13.875 \approx 13.9,$$
$$12.55 \times 1.111 = 13.94305 \approx 13.94,$$
$$12.545 \times 1.1111 = 13.9387495 \approx 13.939.$$

Since $\dfrac{138}{11} \times \dfrac{10}{9} = \dfrac{1380}{99} = 13.\overline{93} = 13.939393\ldots$, our calculated approximations are quite good.

Similarly, for this r and s we calculate 1-, 2-, 3-, 4-, and 5-place approximations to $r \div s$, to obtain:

$$10 \div 1 = 10,$$
$$13 \div 1.1 \approx 12,$$
$$12.5 \div 1.11 \approx 11.3,$$
$$12.55 \div 1.111 \approx 11.30,$$
$$12.545 \div 1.1111 \approx 11.291.$$

Actually $\dfrac{138}{11} \div \dfrac{10}{9} = \dfrac{138}{11} \times \dfrac{9}{10} = \dfrac{1242}{110} = 11.29\overline{09} = 11.290909\ldots$, and hence our approximations are fairly good.

Exercises

Round off each of the following decimals to 1, 2, 3, 4, and 6 *decimal places*, and also round off each of them to 1, 2, 3, and 4 *places*:

1. 507.2856496;

2. 0.001782475;

3. 62.4999876;

4. 0.0002473829.

In each of the Exercises 5 to 8, round off the values of r and s to one decimal place and then get a 1-decimal-place approximation to $r + s$ and $r - s$. Do the same for 2, 3, 4, and 7 decimal places.

5. $r = 28.\overline{7}$, $s = 5.6\overline{48}$;

6. $r = 6.84\overline{3}$, $s = 0.00\overline{59}$;

7. $r = .02\overline{31}$, $s = 0.00\overline{746}$;

8. $r = 0.000\overline{261}$, $s = 86.\overline{15}$;

In each of the Exercises 9 to 12, round off the values of x and y to one place and then get a 1-place approximation to xy and $x \div y$. Do the same for 2, 3, and 5 places.

9. $x = 593.\overline{2}$, $y = 8.\overline{1}$;

10. $x = 0.002\overline{3}$, $y = 0.000\overline{85}$;

11. $x = 37.800.\overline{5}$, $y = 0.004\overline{6}$;

12. $x = 0.6\overline{25}$, $y = 8.\overline{74}$;

9.5   The real numbers

The decimals that we have been considering are called **real numbers** and we formalize this with the following definition.

Definition 9.5.1. A *real number is a number that can be expressed in the form*

$$\pm a_k\, a_{k-1} \cdots a_2\, a_1\, a_0 . a_{-1}\, a_{-2} \cdots$$

where k is an integer and each a is one of the numbers $0, 1, 2, \ldots, 9$.

We shall often use the letter D to denote the set of real numbers. As we have seen, a real number may be a terminating decimal, a repeating decimal, or a nonrepeating decimal. The system D of real numbers, then, contains the system of rational numbers, which includes the system of integers, which, in turn, contains the system of whole numbers. We have thus extended the chain one link farther.

Definition 9.5.2. *Two real numbers r and s are called* **equal** *(written r = s or s = r), if and only if either of the following holds:*

(1) *They are identical; that is, every digit of each is equal to the corresponding digit of the other.*

(2) *One of the numbers is a terminating decimal and the other number can be obtained from it by decreasing the last nonzero digit of the first real number by 1 and changing every digit after this last nonzero one to a 9.*

For example:

$$7.2\overline{8} = 7.2\overline{8},$$

$$6.28 = 6.27999\ldots = 6.27\overline{9},$$

$$320 = 319.99\overline{9} = 319.\overline{9},$$

$$100 = 99.99\overline{9} = 99.\overline{9},$$

$$0.6281 = 0.62809,$$

$$0.51\overline{9} = 0.52.$$

Similarly to the way we treated other kinds of equality in Theorems 3.1.1, 3.8.1, 6.1.1, and 8.1.1, we shall now see that our equality of real numbers satisfies the reflexive, symmetric, and transitive properties, so that it is an **equivalence relation.**

Theorem 9.5.1. *If r, s, and t are real numbers, then:*

(1) $r = r$ *(Equality of real numbers is reflexive);*

(2) *If* $r = s$, *then* $s = r$ *(Equality of real numbers is symmetric);*

(3) *If* $r = s$ *and* $s = t$, *then* $r = t$ *(Equality of real numbers is transitive).*

Proof.

(1) $r = r$ because r is identical to itself.

(2) If $r = s$, then $s = r$ because in the definition of equality the order of r and s is immaterial.

(3) Let $r = s$ and $s = t$. If either r and s are identical or s and t are identical, then it follows immediately from the definition of equality that $r = t$. If neither r and s nor s and t are identical, then one of each of these pairs must be a decimal with repeating 9's and the other must be a terminating decimal. Since s occurs in both pairs, r and t must be identical in this case, and hence equal.

9.6   Addition, multiplication, subtraction, and division of real numbers

In Section 9.4 we considered methods of getting "approximations" to sums, differences, products, and quotients of real numbers. We did this, in an informal way, even though we had not defined real numbers or their operations. In this section we shall use those informal "approximations" to help define addition and multiplication and, indirectly, subtraction and division of real numbers.

In particular, in Section 9.4 we saw how to get an n-decimal-place "approximation" to $r + s$ from n-decimal-place "approximations" to r and s. Using this technique on two real numbers r and s, we can get an n-decimal-place "approximation to $r + s$". It can be proved, using the theory of limits of infinite sequences (a topic beyond the scope of this book), that if n is sufficiently large, this n-decimal-place "approximation" becomes arbitrarily close to a unique real number. We define the sum $r + s$ to be the unique real number thus approached.

The situation for multiplication is similar to that of addition. We also saw in Section 9.4 how to get an n-place "approximation to rs" from n-place "approximations" to r and s. Using this on two real numbers r and s an n-place "approximation to rs" can be obtained. As in the case of addition, it can be proved, using the theory of limits of infinite sequences, that if n is taken sufficiently large, this n-place approximation becomes arbitrarily close to a unique real number. We define the product rs to be the unique real number thus approached.

We formalize these statements about addition and multiplication with the following definitions:

Definition 9.6.1.   *The sum $r + s$ of real numbers r and s is the number approached by the n-decimal-place approximation to $r + s$ as n increases without limit.*

Example 1. $1.6 + 0.\overline{09} = 1.666666\ldots + 0.090909\ldots = 1.757575\ldots.$

Example 2. $3.202002000200002\ldots + 0.003003000300003\ldots$
$$= 3.205005000500005\ldots.$$

Notice that we can find as many decimal places as we wish for each of these sums. In fact we can see that in the first case the sum is $1.\overline{75}$ and in the second case the pattern of how the decimal places continue is quite discernible. Notice that the real numbers in the second example are not rational because they are not repeating decimals.

Definition 9.6.2. *The product rs of real numbers r and s is the number approached by the n-place approximation to rs as n increases without limit.*

For example, using the numbers of Section 9.4, we have

$$12.\overline{54} \times 1.\overline{1} = 12.545454\ldots \times 1.111111\ldots.$$

The 3-place approximation obtained is 13.9, the 5-place one is 13.939, and the 10-place approximation is 13.93939394. We can find as many decimal places in the product as we wish. From the 10 place approximation we might well suspect that the number approached is $13.\overline{93}$, which it is.

There may be no discernible pattern to the decimal places of a product, however, and then we can only approximate the product. Nonetheless, we can make the approximation as near to the product as we wish, provided we spend enough time calculating.

We now define subtraction and division in the expected way.

Definition 9.6.3. *If r and s are real numbers, then r − s is the number x such that*

$$s + x = r.$$

Definition 9.6.4. *If r and s are real numbers, then r ÷ s is the number x such that*

$$sx = r.$$

It can be proved, using the theory of limits, that such unique real numbers do exist, except of course that $r \div 0$ does not exist for any real number r, where $0 = 0.000\ldots = .\overline{0}$. In fact, it can be proved that similarly to addition and multiplication, $r - s$ and $r \div s$ are the unique real numbers approached by their approximations described in Section 9.4.

Naturally we would hope that addition, multiplication, subtraction, and division are **well defined**. This is the case as the following theorem indicates, but again the proof depends on the theory of limits and will not be given.

Theorem 9.6.1. *If r, s, r′, and s′ are real numbers such that r = r′ and s = s′, then*

(1) $r + s = r' + s';$
(2) $rs = r's';$
(3) $r - s = r' - s';$
(4) $r \div s = r' \div s'$ *provided* $s \neq 0.$

Now that subtraction is defined, it is convenient to consider the converse of Theorem 9.2.1. Must a repeating decimal be a rational number? The answer is affirmative. Consider for example the decimal $d = 6.1\overline{23}$. Notice that $100d = 10^2 d = 612.\overline{323}$ is the same as d after the first decimal place, so that

$$100d - d = 612.3\overline{23} - 6.1\overline{23} = 606.2.$$

Therefore

$$100d - d = (100 - 1)d = 99d = 606.2,$$

and we have

$$d = \frac{606.2}{99} = \frac{6062}{990} = \frac{3031}{495}.$$

Hence $d = 6.1\overline{23}$ is the rational number $\dfrac{3031}{495}$.

Similarly if we had started with $d = 87.62\overline{574}$, we would take the equation

$$10^3 d - d = 999d = 87625.74\overline{574} - 87.62\overline{574} = 87538.12$$

Hence

$$d = \frac{87538.12}{999} = \frac{8753812}{99900}$$

We state the general result as the following theorem:

Theorem 9.6.2. *Every repeating decimal is a rational number.*

Proof. If d is a repeating decimal with a sequence of h digits that forever repeat, then $10^h d - d$ is a terminating decimal t (which is a rational number whose denominator is a power of 10). Hence

$$(10^h - 1)d = t.$$

Therefore

$$d = \frac{t}{10^h - 1},$$

which is a rational number.

Like the rational numbers, the system D of real numbers forms what is called a **field**. That is, it has the properties stated in the following theorem. Here, of course, $0 = 0.\overline{0}$ and $1 = 1.\overline{0}$.

Theorem 9.6.3. *If r, s, and t are in D, then*
    (1) $r + s \in D$;   (2) $r + s = s + r$;
    (3) $(r + s) + t = r + (s + t)$;   (4) $r + 0 = r$;
    (5) $r - s \in D$;   (6) $rs \in D$;
    (7) $rs = sr$;   (8) $(rs)t = r(st)$;
    (9) $r \cdot 1 = r$;   (10) if $s \neq 0$, $r \div s \in D$;
  (11) $r(s + t) = rs + rt$; $(s + t)r = sr + tr$.

We will not give a proof for this theorem or Theorem 9.6.5 below because they involve the theory of limits, which, as we have previously mentioned, is beyond the scope of this book. We will point out, however, that the proof of each part depends essentially on the fact that the corresponding result holds for the terminating decimals which are the n-decimal-place or n-place approximations to the sums, differences, products, and quotients involved.

Next we define the negative of any real number r in the same way as we defined the negative of an integer and a rational number.

Definition 9.6.5. *If r is a real number, then the number x such that $r + x = 0$ is called the **negative** of r, and is written as $-r$.*

It should be noted that this definition is consistent with our notation for real numbers of the form

$$-a_k a_{k-1} \cdots a_2 a_1 a_0.a_{-1} a_{-2} \cdots$$

in Definition 9.5.1. In particular, let us add $323.\overline{23} + (-323.\overline{23})$. Now

$$323.\overline{23} = 3(10)^2 + 2(10) + 3 + 2(10)^{-1} + 3(10)^{-2} + \cdots$$

and

$$
\begin{aligned}
-323.\overline{23} &= -[3(10)^2 + 2(10) + 3 + 2(10)^{-1} + 3(10)^{-2} + \cdots] \\
&= -3(10)^2 - 2(10) - 3 - 2(10)^{-1} - 3(10)^{-2} - \cdots .
\end{aligned}
$$

Hence

$$323.\overline{23} + (-323,\overline{23}) = 0(10)^2 + 0(10) + 0 + 0(10)^{-1} + 0(10)^{-2} + \cdots = 0.$$

Similarly, in the general situation, when we add

$$[a_k a_{k-1} \cdots a_2 a_1 a_0.a_{-1} a_{-2} \cdots] + [-a_k a_{k-1} a_2 a_1 a_0.a_{-1} a_{-2} \cdots],$$

we obtain

$$0(10)^k + 0(10)^{k-1} + \cdots + 0(10) + 0 + 0(10)^{-1} + 0(10)^{-2} + \cdots = 0.$$

Remember, however, that although $-323.\overline{23}$ is the negative of $323.\overline{23}$, $323.\overline{23}$ is also the negative of $-323.\overline{23}$, that is, $-(-323.\overline{23}) = 323.\overline{23}$. In general (just as for rational numbers), $-(-r) = r$, because if $r + (-r) = 0$, then $(-r) + r = 0$. Hence, by definition of negative, $-(-r) = r$.

We will define positive and negative real numbers as follows.

Definition 9.6.6. *Let each a be one of the numbers $0, 1, 2, \ldots, 9$. A nonzero number of the form $a_k a_{k-1} \cdots a_2 a_1 a_0.a_{-1} a_{-2} \ldots$ is called a **positive number**. A nonzero number of the form $-a_k a_{k-1} \cdots a_2 a_1 a_0.a_{-1} a_{-2} \cdots$ is called a **negative number**.*

For example, 6.2, 7, 8.3$\overline{6}$, 0.00$\overline{17}$ are positive, while -6.2, -7, $-8.3\overline{6}$, and $-0.00\overline{17}$ are negative. A theorem similar to Theorem 8.5.4 now follows.

Theorem 9.6.4. *Let r be a real number.*
    (1) *If r is a positive number, then $-r$ is a negative number;*
    (2) *If $r = 0$, then $-r = 0$;*
    (3) *If r is a negative number, then $-r$ is a positive number.*

Similar to Theorems 6.6.1 and 8.9.1 is the following theorem which helps when calculating with real numbers, some of which may be negative.

Theorem 9.6.5. *Let r and s be nonnegative real numbers. Then*
    (1) $(-r) + s = s + (-r) = s - r = -(r - s)$;
    (2) $(-r) + (-s) = (-r) - s = (-s) - r = -(r + s)$;
    (3) $(-r) - (-s) = (-r) + s = s - r = -(r - s)$;
    (4) $r - (-s) = r + s$;
    (5) $(-r)s = r(-s) = -(rs)$;
    (6) $(-r)(-s) = rs$;
    (7) *if $s \neq 0$,* $(-r) \div s = r \div (-s) = -(r \div s)$;
    (8) *if $s \neq 0$,* $(-r) \div (-s) = r \div s$.

As with the rational numbers, we shall often write $r \div s$ as $\dfrac{r}{s}$ or r/s. From parts (7) and (8) of Theorem 9.6.5, we then have

$$\frac{-r}{s} = \frac{r}{-s} = -\frac{r}{s} \quad \text{and} \quad \frac{-r}{-s} = \frac{r}{s},$$

for any real numbers r and s with $s \neq 0$.

Other rational number theorems that also hold for real numbers are Theorems 8.4.2 through 8.4.7. In fact, several parts of their proofs are essentially the same as the proofs for rational numbers.

Exercises

1. Write the statements of Theorems 8.4.2 through 8.4.7 for real numbers.
2. Express each of the following repeating decimals in the form a/b where a and b are integers:
    (a) 6.2$\overline{5}$;   (b) 0.00$\overline{75}$;   (c) 0.0631;
    (d) 4.3$\overline{21}$;   (e) 6.1$\overline{723}$;   (f) 0.036$\overline{245}$.
3. Find the sum and product of each of the following pairs of real numbers:
    (a) 2.1$\overline{2}$, 3.$\overline{4}$;   (b) 0.1$\overline{32}$, 6.1;   (c) 0.$\overline{12}$, 0.$\overline{23}$;
    (d) 0.1010010001..., 0.030030003... where the patterns continue.

9.7   Inequalities and absolute values for real numbers

We will define $<$ and $>$ as follows:

Definition 9.7.1. *Let r and s be real numbers. Then r $<$ s and s $>$ r if and only if there exists a positive real number p such that r + p = s.*

The whole theory that we developed for inequalities of integers (Theorems 6.8.2, 6.8.3, 6.8.4, and related material) now can be proved using almost identical proofs.

We also use the same definition of absolute value for real numbers as we used for rational numbers in Definition 8.10.1.

Definition 9.7.2. *The **absolute value** $|r|$ of a real number r is u if u $\geq$ 0 and is* $-u$ *if u $<$ 0.*

For example, $|-3.0\overline{51}| = -(-3.0\overline{51}) = 3.0\overline{51}$.

Exercises

1. Write the statements of Theorems 6.8.2, 6.8.3, and 6.8.4 for real numbers.

9.8   Real numbers on the number line

The rational numbers do not fill up the number line, and we can show that there are points on the line that are not associated with rational numbers. One such point can be located as follows. Take a square, $ABCD$, with sides 1 unit long, and let the diagonal AC be h units in length. Place the diagonal AC of the square on the number line with point A at the origin as in Figure 9.8.1. The point C is then a point on the number line that is not associated with any rational number, as we will now show. We shall use a geometrical theorem, **The Theorem of Pythagoras**, which states that the sum of the squares of the lengths of the two shorter sides of a right triangle is equal to the square of the length of the hypotenuse (the longest side). Thus, for the Figure 9.8.2, $a^2 + b^2 = c^2$.

Applying this to triangle ADC on the number line we get $h^2 = 1^2 + 1^2 = 1 + 1 = 2$. That is, h is a number whose square is 2. We will now show that there is no rational number h such that $h^2 = 2$.

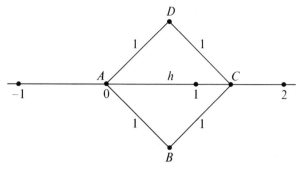

Figure 9.8.1

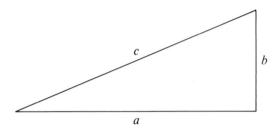

Figure 9.8.2

If there were such a rational number we should have integers n and d such that $h = n/d$, and

$$\left(\frac{n}{d}\right)^2 = 2.$$

We assume that the fraction n/d is in lowest terms so that n and d have no common factor; that is, $(n, d) = 1$. Now

$$\left(\frac{n}{d}\right)^2 = \frac{n^2}{d^2},$$

so we have

$$\frac{n^2}{d^2} = 2.$$

Hence

$$n^2 = 2d^2.$$

But since factorization into primes is unique by the Fundamental Theorem of Arithmetic (Theorem 7.3.3), we must have $2|n$. Therefore, $n = 2k$ where k is an integer. Then,

$$n^2 = (2k)^2 = 4k^2 = 2d^2.$$

Dividing by 2 we have

$$2k^2 = d^2.$$

Now by the unique factorization theorem again, we must have $2|d$. Thus, we have shown that $2|n$ and $2|d$, so n and d have the common factor 2, which is contrary to our assumption that the fraction n/d is in lowest terms. Hence, there is no rational number h such that $h^2 = 2$. By a similar proof, we could show that there is no *rational* number x such that $x^2 = 10$ or 15 or 20 or 3. In fact, it can be proved that there is a *rational* number x such that x^2 is an integer w, only when w is a perfect square of an *integer*; that is, when $w = u^2$ for an integer u. For example, there is an integer, and hence a rational number, x such that $x^2 = 16$, namely $x = 4$, but there is no rational number x such that $x^2 = 20$.

It can be proved, however, that for every positive integer w, there is a positive real number r such that $r^2 = w$. We will write $\sqrt{w}$ for such a number and call it **the square root** of w. A 24-decimal-place approximation to $\sqrt{2}$ is $\sqrt{2} \approx 1.414213562373095048801689$.

Another example of a real number that is not a rational number is the number usually denoted by π. The number π is defined as the ratio of the length c of a circle to the length d of its diameter. (*See* Figure 9.8.3.) The

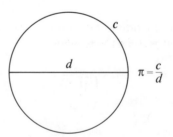

Figure 9.8.3

number π is the same for circles of all sizes. Also, c and d cannot both be rational, because then π would have to be rational. A 15-decimal-place approximation to π is 3.141592653589793.

Real numbers like π and $\sqrt{2}$ that are not rational are called **irrational numbers** according to the following definition.

Definition 9.8.1. *An **irrational number** is a real number that is not a rational number.*

The real numbers, then, are composed of the rational numbers and the irrational numbers. Since a number is rational if and only if it can be expressed as a repeating decimal, the irrational numbers have nonrepeating decimal numerals.

Now let us consider how each real number is associated with a unique point on the number line and how each point on the number line is associated with a uniqe real number.

First let us consider how a particular number, say, $\pi = 3.14159\ldots$, is associated with a point on the number line. We first go to 3.1, which is one-tenth of the way from 3 to 4.

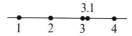

We then divide the interval between 3.1 and 3.2 into ten equal parts. A magnified drawing of this is as follows.

The fourth point to the right of 3.1 is then 3.14. Next we divide the interval from 3.14 to 3.15 into ten equal parts. The first one of these to the right of 3.14 is the point 3.141. We continue this process, finding 3.1415, 3.14159, etc. We get closer and closer—in fact, arbitrarily close—to a unique point which is the point associated with the real number π.

In general, for any real number r, we first find the point on the number line for the rational number with one decimal place obtained by omitting all digits after the first decimal place of r; then we find the point on the number line for the rational number with two decimal places obtained by eliminating all digits after the second decimal place; then we find a point on the line for the rational number with three decimal places obtained by eliminating all digits after the third. It can be proved that as we continue this process we get closer and closer to a unique point P on the number line. This point P is the point on the number line associated with the real number r. In practice, unless we know that we have a number that can be constructed like $\sqrt{2}$, we can only approximate its location on the line. In fact, even when we construct $\sqrt{2}$ we are only approximating its location, our accuracy depending on the accuracy of the construction.

It is also the case that if we take any point P on the number line, there is a unique real number r that we can associate with it. We illustrate this for the point P shown in Figure 9.8.4. In our discussion we shall assume that P lies to the right of 0. If P is to the left of 0, the necessary changes in the discussion will be clear.

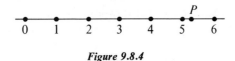

Figure 9.8.4

The point P is either at a point labeled with an integer or between two such points. In the first case the number r is the integer at which P occurs. In the second case, the real number associated with P has the smaller of the two integers between which P occurs as the part to the left of the decimal point. In Figure 9.8.5 this number is 5. The interval between successive integers in which P occurs is then divided into ten equal intervals.

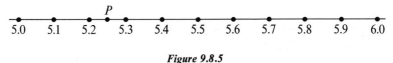

Figure 9.8.5

As before, the point P will either be at one of these division points or between two division points.

In the first case, r is the terminating decimal with one decimal place associated with that point. In the second case, the terminating decimal with one decimal place associated with the nearest division point to the left of P is a one-decimal approximation to r obtained by eliminating all of its decimal places after the first. In our specific illustration, this approximation is 5.2. We then divide this smaller interval into ten equal intervals and continue in this way, finding successively better and better approximations to the unique number r associated with P.

Exercises

1. Prove that $\sqrt{3}$ is not a rational number.
2. Prove that $\sqrt{6}$ is not a rational number.
3. Draw a line 10 inches long and label it with the integers from -5 to 5, using 1-inch intervals. On your figure locate as accurately as you can each of the following real numbers:
   (a) $\sqrt{2}$;   (b) π;   (c) $-\sqrt{2}$;   (d) $-\pi$;   (e) $\pi + \sqrt{2}$;
   (f) $\pi - \sqrt{2}$;   (g) $\sqrt{2} - \pi$;   (h) $-\pi - \sqrt{2}$;   (i) $2.31010010001\ldots$.

9.9   The density and completeness properties

The real numbers have two other important properties that should be mentioned. These are called the **density property** and the **completeness property**.

The density property is the property that if r and s are any two unequal real numbers, there is always at least one real number between them. We state the property formally in the following theorem.

Theorem 9.9.1. (*The density property*) If r and s are real numbers such that $r < s$, then there exists a real number d such that $r < d < s$.

Proof. We will show that $r < (r + s)/2 < s$; that is, that d can be taken to be $(r + s)/2$. From the inequality

$$r < s,$$

by adding first r and then s to both sides we obtain:

$$r + r < r + s \quad \text{and} \quad r + s < s + s.$$

Thus we have

$$2r < r + s < 2s.$$

Dividing by the positive number 2 we get

$$r < \frac{r + s}{2} < s.$$

This completes the proof. Actually the number $(r + s)/2$ is the point on the number line midway between r and s. This same proof holds if r and s are assumed to be rational numbers, $(r + s)/2$ being a rational number in that case. Thus the rational numbers by themselves also have the density property. In fact it can be proved that between any two unequal real numbers there is always a rational number and an irrational number.

In practice it is not difficult to find a terminating decimal (which is therefore rational) between two given unequal real numbers. Consider for example:

$$a = 2.36841492\ldots,$$
$$b = 2.36841571\ldots.$$

The rational number $r = 2.3684152$ is such that $a < r < b$. It is not necessary to know what the decimal places of a and b are beyond those given to find this r. Another choice for r is 2.3684155.

It follows that if a and b are real numbers such that $a < b$, then there are infinitely many rational numbers between a and b. This is because there is a rational number r_1 between a and b; that is, $a < r_1 < b$. But r_1 and b are unequal real numbers, so there is a rational number r_2 between r_1 and b; that is, $a < r_1 < r_2 < b$. Similarly since r_2 and b are unequal real numbers, there is a rational number r_3 such that

$$a < r_1 < r_2 < r_3 < b.$$

By continuing in this way we can insert as many rational numbers, r_1, r_2, r_3, ..., as we wish between a and b; that is, we have

$$a < r_1 < r_2 < r_3 < \cdots < b.$$

Before describing the completeness property, we shall consider two related concepts needed to understand it, namely **upper bound** and **least upper bound** of a set of numbers.

Definition 9.9.1. *If S is a nonempty set of real numbers and b is a real number (which may or may not be in S) such that $x \le b$ for all x in S, then b is called an* **upper bound** *of S.*

Notice that if b is an upper bound for a set S, then any number greater than b is also an upper bound for S.

In the following examples remember that D is the system of real numbers.

Example 1. Let $S = \{x \in D \,|\, x \text{ is negative}\}$. Then 0 is an upper bound for S since $x \le 0$ for all x in S. Note that every positive real number is also an upper bound for S. In particular, 1, 2, and 617 are upper bounds for S. The number 0 however, is the least number which is an upper bound for S.

Example 2. Let $S = \{x \in D \,|\, x^2 < 2\}$. Here 7 is an upper bound for S. Also 16, $1\frac{1}{2}$, and $\sqrt{2}$ are upper bounds for S. Notice that $\sqrt{2}$ is the least number which is an upper bound for S.

Example 3. Let $S = D$. In this case S has no upper bound.

Example 4. Let S be the set of all terminating decimals. Here again S has no upper bound.

These examples also illustrate the concept of a **least upper bound**, which is defined as follows:

Definition 9.9.2. *If S is a nonempty set of real numbers and u is an upper bound for S and S has no upper bound less than u, then u is called the* **least upper bound** *of S.*

Analogous to the notions of upper bound and least upper bound are those of **lower bound** and **greatest lower bound**. The statements of the definitions of these are left as an exercise.

The following property is called the **completeness property** of the system of real numbers. Its proof is beyond the scope of this book.

Theorem 9.9.2. *If a nonempty set of real numbers is bounded above, it has a least upper bound; if a nonempty set of real numbers is bounded below, it has a greatest lower bound.*

The significance of the word "completeness" in the completeness property for real numbers is that the least upper bound and greatest lower bound mentioned in Theorem 9.9.2 are not outside the system of real numbers. Now the system of rational numbers is *not complete* in the sense that when a nonempty set of rational numbers is bounded above, it has a least upper bound, but the least upper bound does not have to be a rational number. It may be outside the system of rational numbers. For example, the set $S = \{x \in F \mid x^2 < 2\}$ has a rational upper bound, say 3, but there is no rational number u that is a least upper bound for S. As we have mentioned above however, the irrational number $\sqrt{2}$ is the least upper bound for this set S. Similarly, the greatest lower bound of a set of rational numbers that has a lower bound may not be a rational number. The system of rational numbers does not have the completeness property.

Exercises

1. Find three rational numbers between the numbers of each of the following pairs of real numbers.
   (a) $0.136824\ldots$ and $0.136791\ldots$; (b) $6.1830001\ldots$ and $6.182\overline{9}$;
   (c) $-0.476257\ldots$ and $-0.4705841\ldots$; (d) $-1.826\ldots$ and $-1.8252\ldots$.

2. Find 3 upper bounds and the least upper bound, if they exist, for each of the following sets S of numbers. If the set is a set of rational numbers and has a least upper bound, state whether the least upper bound is rational. (F denotes the system of rational numbers, and D denotes the system of real numbers.)
   (a) $S = \{x \in F \mid x < 5\}$; (b) $S = \{x \in F \mid x^2 < 3\}$;
   (c) $S = \{x \in D \mid x^2 < 3\}$; (d) $S = F$;
   (e) $S = \{x \in F \mid x < -\pi\}$; (f) $S = D$.

3. State definitions analogous to Definitions 9.9.1 and 9.9.2 for the notions of lower bound and greatest lower bound.

ANSWERS TO CERTAIN EXERCISES

These answers can be very valuable if used properly. After working out the first exercise in an assignment, a student should refer to the first answer. If it checks he is probably doing the exercises correctly; but if it does not check, he should go back over his problem and try to find the error. He should proceed with this same technique until the assignment is complete. The value in referring to the answer after doing each problem is that a student is less likely to do more than one problem using the wrong technique. It is usually a poor policy to work a problem by referring to the answer before finding one, although in instances where a student is really stuck this may be of value in getting started.

Section 1.2
1. F.　2. F.　3. F.　4. F.　5. F. (Some people who are now babies will, at some future time, marry people who have not been born yet.)　6. F.　7. T.　8. T.　9. T. (The unique number is 1.)　10. T.　11. T.　12. T.　13. T.

Section 1.4
1. It shows that they have the same truth value for some cases.　3. Mars is not a planet or some cows are not black. Statement F, negation T.　4. Mars is not a planet and some cows are not black. Statement T, negation F.　5. All cows are black or Mars is not a planet. Statement T, negation F.　6. Bill is not wearing a red tie or George is not wearing a tie.　7. Bill is wearing a tie and Jim is not tired.　8. No people are hot and some cows are not black. Statement T, negation F.　9. Some cows are white or no people are simple. Statement F, negation T.　10. All cows are tall or some people are wide. Statement F, negation T.　11. All cows are simple or all people are problems. Statement T, negation F.　12. All corn is sweet and all cows are problems. Statement T, negation F.

Section 2.1
1. invalid.　2. valid.　3. (a) invalid　(b) invalid　(c) invalid　(d) valid　(e) invalid.　4. (a) invalid　(b) valid.　5. (a) invalid　(b) valid　(c) invalid. Hypothesis T, (a) T　(b) T　(c) T.　6. (a) valid　(b) invalid　(c) invalid　(d) invalid. Hypothesis F, (a) T　(b) T　(c) F　(d) F.

Section 2.2

1. F.   **2.** T.   **3.** T.   **4.** T.   **5.** T.   **6.** T.   **7.** F.   **8.** F.   **9.** (c).
10. (c).   **11.** (a).   **12.** (a) If Jim goes, then Bill will go.   (b) If Bill does not go, then Jim will not go.   (c) If Jim does not go, then Bill will not go.   **13.** (a) If Bill does not go, then Jim will not go.   (b) If Jim goes then Bill will go. (c) If Bill goes then Jim will go.   **14.** (a) If Bill goes then Jim will go.   (b) If Jim does not go, then Bill will not go.   (c) If Bill does not go, then Jim will not go.   **15.** (a) If $y = 7$ then $x = 2$.   (b) If $x \neq 2$ then $y \neq 7$.   (c) If $y \neq 7$ then $x \neq 2$.   **16.** (a) If $y \neq 7$ then $x \neq 2$.   (b) If $x = 2$ then $y = 7$.   (c) If $y = 7$ then $x = 2$.   **17.** (a) If $y \neq 5$ then $x = 3$.   (b) If $x \neq 3$ then $y = 5$.   (c) If $y = 5$ then $x \neq 3$.   **18.** Given implication T.   (a) If $x = 2$ then $2x = 4$, T.   (b) If $2x \neq 4$ then $x \neq 2$, T.   (c) If $x \neq 2$ then $2x \neq 4$, T.   **19.** Given implication T.   (a) If $2x \neq 4$ then $x \neq 2$, T.   (b) If $x = 2$ then $2x = 4$, T.   (c) If $2x = 4$ then $x = 2$, T.   **20.** Given implication F.   (a) If $x = 2$ and $y = 6$ then $x + y = 8$, T.   (b) If $x + y \neq 8$ then $x \neq 2$ or $y \neq 6$, T.   (c) If $x \neq 2$ or $y \neq 6$ then $x + y \neq 8$, F.

Section 2.3

1. Proceeding as in the hint, we assume that Jim's tie is not red. Case (1). Bill's tie is not red. In this case, George will see that neither Bill nor Jim has a red tie and since he knows there is at least one red tie, he will claim to have a red tie. Case (2). Bill's tie is red. In this case, George will not say anything because he sees a red tie and Bill will realize this and thus know about his red tie, and hence he will claim to have a red tie.

We have now proved that if neither Bill nor George claims to have a red tie, then Jim has a red tie, by proving its contrapositive. Thus Jim would know he has a red tie if no one speaks for several minutes.

Section 3.3

1. (a) $\varnothing$ (b) $\varnothing$, $\{7\}$ (c) $\varnothing$, $\{1\}$, $\{2\}$, $\{1, 2\}$ (d) $\varnothing$, $\{a\}$, $\{b\}$, $\{c\}$, $\{a, b\}$, $\{a, c\}$, $\{b, c\}$, $\{a, b, c\}$ (e) $\varnothing$, $\{a\}$, $\{\{b, c\}\}$, $\{a, \{b, c\}\}$.   **2.** Because every element of each set is an element of the other.   **3.** (a) The set of whole numbers greater than 1.   (b) $\{13, 14, 15, 16, 17, 18, 19\}$.   (c) The set whose elements are the numbers 2, 3, 4 and the blue-eyed cats owned by people named Egbert. (d) $\varnothing$ (e) S_{11} (f) $\varnothing$.   **4.** (a) $\{b, c\}$ (b) $\{a, b, c, d, e\}$ (c) $\{b, d\}$ (d) $\{a, b, d\}$ (e) T.   (f) $\{b\}$.   **5.** No, because $0 \in \{0\}$ but $0 \notin \varnothing$.   **6.** Yes, because $\varnothing$ is a subset of every set.   **7.** Because the elements of $\{a, b, c\}$ are a, b, and c and $\{a\}$ is not one of them.   **8.** Because $\{a\}$ is one of the elements of the set.   **9.** $S \cap T = T \cap S$ because the elements that are in both S and T are the elements that are in both T and S. $S \cup T = T \cup S$, because the elements that are in S or in T are the elements that are in T or in S.   **10.** (a) S_8 (b) S_8.

Section 3.5

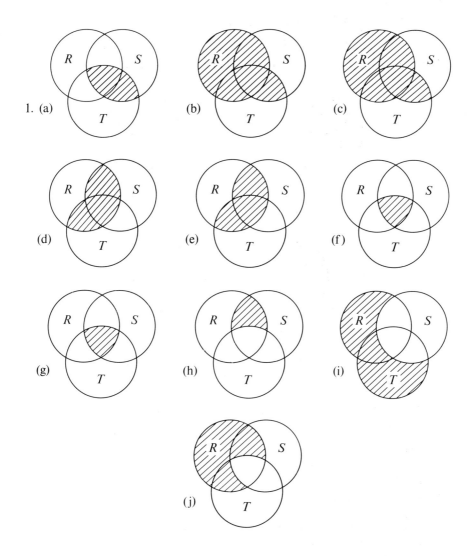

1. (a) (b) (c) (d) (e) (f) (g) (h) (i) (j)

2. (a) {3, 4} (b) {7, 8, 9, 10, 11} (c) {1, 2, 7, 8, 9, 10, ...} (d) ∅ (e) *T*
(f) *S* (g) {7, 8, 9, 10, 11} (h) {1, 2, 7, 8, 9, 10, ...}   **3.** (a) ∅ (b) {x ∈ T | x is
less than 21 years old} (c) {x ∈ T | x is tall and wears glasses} (d) {x ∈ S | 16 <
x < 93} or {x ∈ S | 17 ≤ x ≤ 92} (e) {x ∈ S | x is even} or {x ∈ S | x = 2n for n ∈ S}
(f) {x ∈ S | x = 5n for n ∈ S} (g) {x ∈ S | x > 18}.

Section 3.6
Proof of (2)

$$(R \cup S) \cup T = \{x \mid x \in R \cup S \text{ or } x \in T\} = \{x \mid x \in R \text{ or } x \in S \text{ or } x \in T\}.$$
$$R \cup (S \cup T) = \{x \mid x \in R \text{ or } x \in S \cup T\} = \{x \mid x \in R \text{ or } x \in S \text{ or } x \in T\}.$$

Therefore $(R \cup S) \cup T = R \cup (S \cup T)$. Significant Euler diagrams:

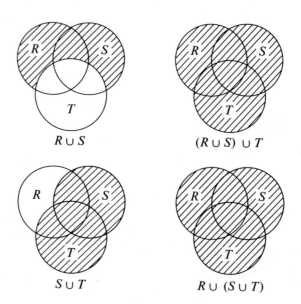

$$R \cup S \qquad\qquad (R \cup S) \cup T$$

$$S \cup T \qquad\qquad R \cup (S \cup T)$$

Proof of (8)
By (5) $(S \cup T) \cap R = R \cap (S \cup T)$, and so by (7) and (5),

$$(S \cup T) \cap R = R \cap (S \cup T) = (R \cap S) \cup (R \cap T) = (S \cap R) \cup (T \cap R).$$

Therefore, $(S \cup T) \cap R = (S \cap R) \cup (T \cap R)$.

Section 3.7
1. $\{(g, 1), (g, 2), (g, 3), (p, 1), (p, 2), (p, 3)\}$.   **2.** $\{(1, g), (2, g), (3, g), (1, p), (2, p), (3, p)\}$.   **3.** $\{(h, g), (h, p)\}$.   **4.** $\{(1, a), (1, b), (1, c), (1, d), (2, a), (2, b), (2, c), (2, d), (3, a), (3, b), (3, c), (3, d)\}$.   **5.** $\{(a, 1), (a, 2), (a, 3), (b, 1), (b, 2), (b, 3), (c, 1), (c, 2), (c, 3), (d, 1), (d, 2), (d, 3)\}$.   **6.** $\{((1, g), h), ((1, p), h), ((2, g), h), ((2, p), h), ((3, g), h), ((3, p), h)\}$.   **7.** $\{(1, (g, h)), (1, (p, h)), (2, (g, h)), (2, (p, h)), (3, (g, h)), (3, (p, h))\}$.   **8.** $\{(1, 1), (1, 2), (1, 3), (2, 1), (2, 2), (2, 3), (3, 1), (3, 2), (3, 3)\}$.   **9.** $\{(g, g), (g, p), (p, p), (p, g)\}$

Section 3.8
1. Two of the six 1–1 correspondence are (a) $a \leftrightarrow p$, $b \leftrightarrow q$, $c \leftrightarrow r$. (b) $a \leftrightarrow q$, $b \leftrightarrow r$, $c \leftrightarrow p$.   **2.** The element $117 \in S$ is paired with $234 \in T$. The element 196 of S is paired with $392 \in T$.   **3.** The point L' is the point where the line OL

crosses the line $A'B'$. The point D of AB is the point where the line OD' crosses the line AB. **4.** $1 \leftrightarrow 101, 2 \leftrightarrow 102, \ldots, n \leftrightarrow 100 + n, \ldots$. That is the 1–1 correspondence is $n \leftrightarrow 100 + n$ for all $n \in S$. n of S is paired with $100 + n$ of T. **5.** (a) $(1, a) \leftrightarrow (a, 1)$, $(1, b) \leftrightarrow (b, 1)$, $(1, c) \leftrightarrow (c, 1)$, $(2, a) \leftrightarrow (a, 2)$, $(2, b) \leftrightarrow (b, 2)$, $(2, c) \leftrightarrow (c, 2)$. **7.** (a) $a \leftrightarrow k, b \leftrightarrow l, c \leftrightarrow m$. (b) $1 \leftrightarrow 6, 2 \leftrightarrow 7, 3 \leftrightarrow 8, 4 \leftrightarrow 9$. (c) $a \leftrightarrow k, b \leftrightarrow l, c \leftrightarrow m, 1 \leftrightarrow 6, 2 \leftrightarrow 7, 3 \leftrightarrow 8, 4 \leftrightarrow 9$. **8.** (a) $1 \leftrightarrow 2, 2 \leftrightarrow 3$. (b) $1 \leftrightarrow 2, 3 \leftrightarrow 3, 4 \leftrightarrow 4$, (c) $(1, 1) \leftrightarrow (2, 2)$, $(1, 3) \leftrightarrow (2, 3)$, $(1, 4) \leftrightarrow (2, 4)$, $(2, 1) \leftrightarrow (3, 2)$, $(2, 3) \leftrightarrow (3, 3)$, $(2, 4) \leftrightarrow (3, 4)$.

Section 4.1

1. $5 = n\{0, 1, 2, 3, 4\}, 6 = n\{0, 1, 2, 3, 4, 5\}$, etc. $C_5 = \{1, 2, 3, 4, 5\}, C_6 = \{1, 2, 3, 4, 5, 6\}$, etc. **2.** $n(C_5) = 5$ since $0 \leftrightarrow 1, 1 \leftrightarrow 2, 2 \leftrightarrow 3, 3 \leftrightarrow 4, 4 \leftrightarrow 5$ is a 1–1 correspondence between $\{0, 1, 2, 3, 4,\}$ and $\{1, 2, 3, 4, 5\}$, etc. **3.** $\{a, b, c, d, e, f, g\} \sim \{1, 2, 3, 4, 5, 6, 7\}$ since $a \leftrightarrow 1, b \leftrightarrow 2, c \leftrightarrow 3, d \leftrightarrow 4, e \leftrightarrow 5, f \leftrightarrow 6, g \leftrightarrow 7$ is a 1–1 correspondence. Therefore, $n\{a, b, c, d, e, f, g\} = 7$. **4.** $\{0, 1, 2\}$ and $\{1, 2, 3\}$ are sets which are not equal. $n\{0, 1, 2\}$ is not the set $\{0, 1, 2\}$, it is the number of elements in the set $\{0, 1, 2\}$, namely 3. **5.** No, because they are both 3. **6.** The number of elements in the set $\{0, 1, 2\}$ is 3, but 3 is not the set itself, that is $3 = n\{0, 1, 2\}$ but $3 \neq \{0, 1, 2\}$.

Section 4.2

1. Let $S = \{0\}$ and $T = C_1 = \{1\}$. Then $S \cap T = \varnothing$. Also, $n(S) = 1$ by definition of 1 and $n(T) = 1$ since T is the counting set C_1. Then $1 + 1 = n(S) + n(T)$. Since $S \cap T = \varnothing$, by definition of addition $n(S) + n(T) = n(S \cup T)$. Therefore, $1 + 1 = n(S) + n(T) = n(S \cup T) = n\{0, 1\}$. But $n\{0, 1\} = 2$ by definition of 2, and therefore $1 + 1 = 2$. **3.** Let $S = C_3 = \{1, 2, 3\}$ and $T = \{a, b, c, d\}$. Then $S \cap T = \varnothing$. Also $n(S) = 3$ since S is the counting set C_3 and $n(T) = 4$ since $T \sim C_4$. Since $S \cap T = \varnothing$, by definition of addition $n(S) + n(T) = n(S \cup T)$. Therefore, $3 + 4 = n(S) + n(T) = n(S \cup T) = n\{1, 2, 3, a, b, c, d\}$. But $\{1, 2, 3, a, b, c, d\} \sim C_7$, so $n\{1, 2, 3, a, b, c, d\} = 7$. Therefore, $3 + 4 = 7$. **5.** Let $S = C_2 = \{1, 2\}$ and $T = C_2 = \{1, 2\}$. Then by definition of multiplication $2 \times 2 = n(S) \times n(T) = n(S \times T) = n\{(1, 1), (1, 2), (2, 1), (2, 2)\}$. But $\{(1, 1), (1, 2), (2, 1), (2, 2)\} \sim C_4$ so $2 \times 2 = 4$. **7.** Let $S = C_4 = \{1, 2, 3, 4\}$ and $T = \varnothing$. Then $n(S) = 4$ and $n(T) = 0$. By definition of multiplication $4 \times 0 = n(S) \times n(T) = n(S \times T) = n(S) \times \varnothing) = n(\varnothing) = 0$. Since $S \times \varnothing = \varnothing$ by Theorem 3.7.1. Therefore $4 \times 0 = 0$. **8.** $n(S \cup T) = 7$, $n(S \cap T) = 2$, $n(S) = 5$, $n(T) = 4$, and $7 + 2 = 5 + 4 = 9$.

Section 4.3

1. (a) yes (b) no (c) yes (d) no (e) yes (f) yes. **2.** (a) yes (b) yes (c) yes (d) no (e) yes (f) yes. **3.** (a) By the commutative property of addition, $(p + q) + r = (q + p) + r$. But by the associative property of addition, $(q + p) + r = q + (p + r)$. Therefore, $(p + q) + r = q + (p + r)$. (b) Using the distributive property twice, and then the associative property of addition, we have $p((q + r) + s) = p(q + r) + ps = (pq + pr) + ps = pq + (pr + ps)$. Therefore, $p((q + r) + s) = pq + (pr + ps)$. **4.** Since $h + p = p$ for all $p \in W$, then in particular $h + p = p$ when $p = 0$. Thus $h + 0 = 0$. But by part (4) of Theorem 4.3.2, $h + 0 = h$. Therefore, $0 = h + 0 = h$ and we have $h = 0$.

Section 4.4

1. (a)70   (b) 46.   **2.** $(a+b)(c+d)=(a+b)c+(a+b)d=ac+bc+ad+bd$.
3. $(a+b)(a+b)=(a+b)a+(a+b)b=aa+ba+ab+bb=aa+2ab+bb$.

Section 4.5

1. 27.

2. (a)    7

(b)    12

(c)    11

(d) 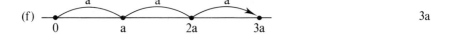   30

(e) 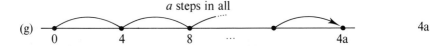   a

(f) 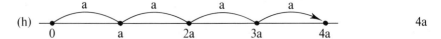   3a

(g)    4a

(h) 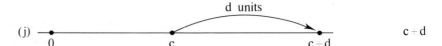   4a

(i)   cd

(j)   c + d

Section 4.6

1. (a) 81   (b) 64   (c) $a^{28}b^{42}$ (d) 1   (e) a^{13} (f) a^{35} (g) 8   (h) $a^{12}c^{20}d^{24}$.
2. (a) 32 (b) 6561 (c) h^{16} (d) r^{h+k} (e) d^{88} (f) $x^{14}y^{21}z^{7}$.   **3.** Using Definition 4.6.2 frequently, we have: (a) $a^{0} \cdot a^{s}=1 \cdot a^{s}=a^{s}$, while $a^{0+s}=a^{s}$, therefore

$a^0 \cdot a^s = a^{0+s}$. (b) $(a^0)^s = 1^s = 1$, while $a^{0 \cdot s} = a^0 = 1$, therefore $(a^0)^s = a^{0 \cdot s}$.
(c) $(ab)^0 = 1$, while $a^0 b^0 = 1 \cdot 1 = 1$, therefore $(ab)^0 = a^0 b^0$. (d) $a^r \cdot a^0 = a^r \cdot 1 = a^r$ while $a^{r+0} = a^r$, therefore $a^r \cdot a^0 = a^{r+0}$. (e) $(a^r)^0 = 1$, while $a^{r \cdot 0} = a^0 = 1$, therefore $(a^r)^0 = a^{r \cdot 0}$.

Section 4.7
1. If $ah = 0$ and $h \neq 0$, then since $0 \cdot h = 0$ we have $ah = 0 \cdot h$. By the restatement we cancel the h to get $a = 0$. **2.** (a) Since p is positive and $p + p = 2p$, then by definition of "$<$", $p < 2p$. (b) Since $5p + 2p = 7p$ and $2p$ is positive, $5p < 7p$.
3. By the distributive property $a(b + 1) = ab + a$. Hence, when a is positive, $ab < a(b + 1)$. **4.** (a) If $a < b$, then by Theorem 4.7.4, since a is positive, $a^2 < ab$. (b) If $a < b$ then by Theorem 4.7.4, since b is positive, $ab < b^2$. But by (a) $a^2 < ab$ and, hence, by Theorem 4.7.2, $a^2 < b^2$.

Section 4.8
1. (b) 2   (c) 13   (d) 0   (f) 9   (h) 2   (i) 0   (k) 3   (m) 9   (n) 5

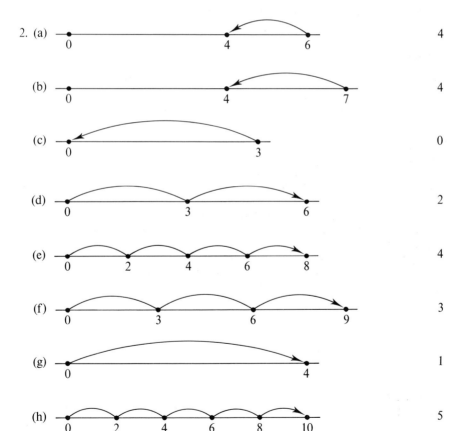

2. (a)   4

(b)   4

(c)   0

(d)   2

(e)   4

(f)   3

(g)   1

(h)   5

(o) 3   (p) 12. The others do not exist.   3. $7|0$ since $7 \cdot 0 = 0$, $0|0$ since $3 \cdot 0 = 0$, or because $a \cdot 0 = 0$ for any a in W, $4 \nmid 10$.   4. If $b|a$ then by definition of divides, there exists a whole number h such that $a = bh$. Therefore, $ac = bhc$ and, hence, $b|ac$.   5. By definition of divides, if $d|a$ and $d|b$, then there exist whole numbers h and k such that $a = dh$ and $b = dk$. Therefore, $a + b = dh + dk = d(h + k)$ and so by definition of divides, $d|(a + b)$.   6. $d = 6$, $a = 4$, $b = 9$. There are infinitely many others.   7. (a) Since by definition of subtraction, $a - b$ is the number which when added to b produces a, $b + (a - b) = a$.   (b) By definition of subtraction, $(a + b) - a$ is the number which when added to a produces $a + b$, hence $(a + b) - a = b$.   8. (a) Since, by definition of division, $a \div b$ is the number which when multiplied by b produces a, $b(a \div b) = a$.   (b) By definition of division, $(ab) \div a$ is the number which when multiplied by a produces ab, and hence $(ab) \div a = b$.

Section 4.9
1. $q = 21$, $r = 3$.   2. $q = 0$, $r = 2$.   3. $q = 0$, $r = 0$.   4. $q = 4$, $r = 5$.
5. $q = 0$, $r = 17$.   6. $q = 89$, $r = 1$.   7. $57 = 5(11) + 2$ and $11 = 5(2) + 1$, so $57 = 5[5(2) + 1] + 2 = 5^2(2) + 5(1) + 2$, so $a_2 = 2$, $a_1 = 1$, $a_0 = 2$.

Section 5.1
1. 123.   2. 367.   3. 54.   4. 1873.   5. 2943.   7. (a)   (b)

(c) XXVI.   9. (a)   (b)   (c) XCIX.

Section 5.2
1. For convenience the symbols $(\ )_5$, $(\ )_3$, and $(\ )_{12}$ are omitted. (a) 0, 1, 2, 3, 4, 10, 11, 12, 13, 14, 20, 21, 22, 23, 24, 30, 31, 32, 33, 34, 40, 41, 42, 43, 44, 100, 101, 102, 103, 104, 110, 111, 112. (b) 0, 1, 2, 10, 11, 12, 20, 21, 22, 100, 101, 102, 110, 111, 112, 120, 121, 122, 200, 201, 202, 210, 211, 212, 220, 221, 222, 1000, 1001, 1002, 1010, 1011, 1012. (c) 0, 1, 2, 3, 4, 5, 6, 7, 8, 9, T, E, 10, 11, 12, 13, 14, 15, 16, 17, 18, 19, 1T, 1E, 20, 21, 22, 23, 24, 25, 26, 27, 28.   2. (a) $(1424)_5$, $(11121)_5$ (b) $(22212)_3$, $(1002010)_3$ (c) $(17E)_{12}$, $(556)_{12}$.   3. 542.   4. 140.   5. $(1033)_4$.
6. 10187.   7. (a) $(11)_5$ (b) $(22)_5$ (c) $(11)_5$ (d) $(13)_5$ (e) $(10)_5$ (f) $(31)_5$ (g) $(11)_3$ (h) $(11)_3$ (i) $(10)_{12}$ (j) $(14)_{12}$ (k) $(12)_{12}$ (l) $(19)_{12}$.

Section 5.3
1. $(22242)_5$.   2. $(13140)_5$.   3. $(102111)_3$.   4. $(21201)_3$.   5. 82052.   6. 12431.

Section 5.4
1. (a) $(6)_7$ (b) $(4)_7$ (c) $(2)_7$.   2. 1639.   3. 4760.   4. $(122)_5$.   5. $(4314)_5$.
6. $(2002)_3$.   7. $(2112)_3$.

Section 5.5
1. 377441.   2. 270648.   3. $(3323114)_5$.   4. $(101011)_5$.   5. $(11011)_3$.
6. $(200022)_3$.

Section 5.6
1. 126.   2. not in W (quotient 506 remainder 835).   3. $(341)_5$.   4. $(213)_5$.
5. $(112)_3$.   6. $(122)_3$.

Section 6.1
1. Because $8 + 0 = 2 + 6$ and $3 + 4 = 7 + 0$, and hence they are equal by the definition of equality in I.   2. No, because $2 + 5 \neq 1 + 3$.   3. $(a + h, h) = (a, 0)$ by definition of equality in I because $a + h + 0 = h + a$; similarly, $(7 + h, 2 + h) = (5, 0)$ because $7 + h + 0 = 2 + h + 5$.   4. (a) $(9, 4)$ (b) $(6, 8)$ (c) $(3, 1)$.   5. (a) $(17, 23)$ (b) $(15, 30)$ (c) $(0, 0)$ (d) $(14, 2)$ (e) $(6, 0)$ (f) $(10, 0)$ (g) $(0, 6)$ (h) $(0, 30)$.   6. (a) $(6, 7) = (9, 10)$ (b) $(18, 24) = (27, 33)$.

Section 6.2
1. (a) is   (b) is not.   2. (a) is not   (b) is not.   3. (a) is   (b) is not.
4. (a) is not   (b) is not.   5. (a) is not   (b) is not.

Section 6.3
1. In I: (1) I is closed under addition,   (2) addition is commutative,   (3) addition is associative,   (4) 0 is the identity of addition,   (5) I is closed under subtraction,   (6) I is closed under multiplication,   (7) multiplication is commutative,   (8) multiplication is associative,   (9) 1 is the identity of multiplication,   (10) multiplication is distributive on both sides over addition.   2–5. Let $u = (a, b)$, $v = (c, d)$ and $w = (e, f)$ where a, b, c, d, e, f are in W.   2. $uv = (a, b)(c, d) = (ac + bd, ad + bc)$ by definition of multiplication in I. But $ac + bd$ and $ad + bc$ are in W, since W is closed under addition and multiplication. Therefore, uv is in I.   3. $(uv)w = [(a, b)(c, d)]w = (ac + bd, ad + bc)(e, f) = (ace + bde + adf + bcf, acf + bdf + ade + bce)$ and $u(vw) = (a, b)[(c, d)(e, f)] = (a, b)(ce + df, cf + de) = (ace + adf + bcf + bde, acf + ade + bce + bdf)$. Thus $(uv)w$ and $u(vw)$ are equal to the same thing and hence, by the transitive property of equality in I, are equal to each other.   4. $u \cdot 1 = (a, b)(1, 0) = (a \cdot 1 + b \cdot 0, a \cdot 0 + b \cdot 1) = (a + 0, 0 + b) = (a, b) = u$. Therefore $u \cdot 1 = u$.

Section 6.4
1. (a) -3 (b) -17 (c) 0   (d) 7   (e) 86   (f) 15.   2. (a) 5   (b) 8   (c) 0   (d) -2 (e) -5 (f) -67.   3. (a) 5   (b) -2 (c) -5 (d) -67 (e) -25 (f) -8.   4. By definition of $-u$, $u + (-u) = 0$. But addition is commutative in I, so $u + (-u) = (-u) + u$. Therefore $(-u) + u = 0$ and, hence, by the definition of the negative of $-u$, $-(-u) = u$.   5. By Theorem 6.4.1, $-u = -(a, b) = (b, a)$ and $-(-u) = -(b, a) = (a, b) = u$. Therefore $-(-u) = u$.

Section 6.5

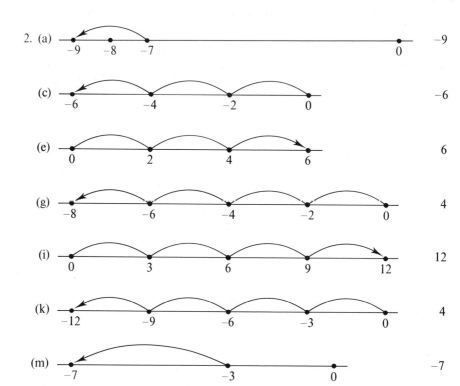

2. (a) ... −9 −8 −7 ... 0 ... −9

(c) ... −6 −4 −2 0 ... −6

(e) ... 0 2 4 6 ... 6

(g) ... −8 −6 −4 −2 0 ... 4

(i) ... 0 3 6 9 12 ... 12

(k) ... −12 −9 −6 −3 0 ... 4

(m) ... −7 −3 0 ... −7

Section 6.6
1. (a) positive   (b) zero   (c) negative   (d) negative   (e) positive   (f) zero.
2. Same as the proof of part (1) in the book, but with the letters a and b interchanged.
3. -6.   **4.** 6.   **5.** -16.   **6.** -13.   **7.** 4.   **8.** -7.   **9.** -7.   **10.** 6.
11. -54.   **12.** 14.   **13.** -107.   **14.** -88.   **15.** -3.   **16.** 12.

Section 6.7
1. $uw + u(v - w) = u[w + (v - w)]$ by the distributive property of multiplication over addition in I. But by the definition of subtraction, $w + (v - w) = v$. Therefore, $uw + u(v - w) = uv$, and thus by the definition of subtraction, $u(v - w) = uv - uw$. To show that $(v - w)u = vu - wu$, apply the commutative property of multiplication to $u(v - w) = uv - uw$.   **2.** (a) $10 - (7 - 1) = 4 \neq (10 - 7) - 1 = 2$.   (b) $12 \div (6 \div 2) = 4 \neq (12 \div 6) \div 2 = 1$.   **3.** (a) $((u \div w) + (v \div w))w = (u \div w)w + (v \div w)w$ by the distributive property, and by the definition of division, $(u \div w)w = u$ and $(v \div w)w = v$. Therefore, $((u \div w) + (v \div w))w = u + v$. Hence, by the definition of division, $(u + v) \div w = (u \div w) + (v \div w)$.   **4.** (a) $12 \div (2 + 4) = 2 \neq (12 \div 2) + (12 \div 4) = 6 + 3 = 9$.   (b) $12 \div (4 - 2) = 6 \neq (12 \div 4) - (12 \div 2) = 3 - 6 = -3$.

Section 6.8

1. By part (2), $x - 4 < 6$ if and only if $x - 4 + 4 < 6 + 4$, that is, if and only if $x < 10$. **3.** $x - 2 > 5$ is equivalent to $5 < x - 2$ and by part (2) $5 < x - 2$ if and only if $5 + 2 < x - 2 + 2$, that is, if and only if $7 < x$, that is, if and only if $x > 7$. **5.** By part (4), $x < 3$ if and only if $5x < 5 \cdot 3$, that is if and only if $5x < 15$. **7.** This is equivalent to $x < -5$ if and only if $2x < -10$. By part (4), $x < -5$ if and only if $2x < 2(-5) = -10$. Thus $2x < -10$ if and only if $x < -5$. **9.** By part (5), $x < -3$ if and only if $-3x > (-3)(-3) = 9$. So $-3x > 9$ if and only if $x < -3$. **11.** $x < 3$ if and only if $-4x > (-4)3 = -12$ by part (5), so $-4x > -12$ if and only if $x < 3$. **12.** There are infinitely many numbers that x can be in each of these; one example for each is (a) -2 (b) -2 (c) 3 (d) -6 (e) 1 (f) -5. **13.** (a) >4 (b) <-2 (c) <-4 (d) >4 (e) >-2 (f) >-6 (g) <24 (h) <-14.

Section 7.1

1. Primes 19, 23, 29, 31, 37 and there are infinitely many others. Composites 60, 63, 65, 66, 69 and there are infinitely many others. **2.** (a) composite (b) composite $(13 \cdot 7)$ (c) prime (d) composite. **3.** Primes 18, 22, 26, 30, 34. Composites 28, 32, 36, 40, 44. **4.** no divisors, prime; $+2, \pm 4, \pm 8$, composite; no divisors, prime; $\pm 2, \pm 10$, composite; no divisors, prime; $\pm 2, \pm 4, \pm 6, \pm 12$, composite; no divisors, prime. **5.** (b) primes $-5, -2, 7, 13, 19$; composites $-14, -8, 4, 10, 16$. **6.** All even integers greater than 2 are composite and, hence, there are infinitely many composites in I.

Section 7.2

1. $a = a \cdot 1$ and $0 = a \cdot 0$ so $a|a$ and $a|0$. Also every common divisor of a and 0 divides a. Therefore, by definition, $g(a, 0) = a$. **2.** $g(36, 76) = 4$. **3.** $g(299, 247) = 13$. **4.** $g(1001, 119) = 7$. **5.** $2^2 \cdot 3$. **6.** $2^3 \cdot 3^2$. **7.** $2^4 \cdot 3^3$. **8.** $2 \cdot 7^2$. **9.** $2 \cdot 5 \cdot 101$. **10.** $7 \cdot 13$.

Section 7.3

1. 9 and 648. **2.** 24 and 1,008. **3.** 18,000 and 72,000. **4.** 60 and $2^4 \cdot 3^3 \cdot 5^2 \cdot 7^3$. **5.** $2^4 \cdot 3^5 \cdot 5^2 \cdot 7 \cdot 11$ and $2^5 \cdot 3^6 \cdot 5^4 \cdot 7^3 \cdot 11^2$. **6.** 2520. **7.** 7,000. **8.** 3360. **9.** 7200. **10.** 3360 **11.** (a) 2380 (b) 616 **12.** $60 = 6 \cdot 10 = 30 \cdot 2$. They are primes, because they have no divisors in E.

Section 7.4

1. 3, 5 only. **2.** none of them. **3.** 3, 11 only. **4.** 3, 9, 11 only. **5.** none of them. **12.** checks. **13.** error, correct answer 36,378. **14.** error, correct answer 2625. **15.** error, correct answer 120,937. **16.** checks. **17.** error, correct answer 621. **18.** error, correct answer $(234)_5$. **19.** error, correct answer $(310222)_5$. **20.** error, correct answer $(1240)_5$.

Section 8.1

1. $\dfrac{3}{2} \neq \dfrac{5}{3}$ because $3 \times 3 = 9 \neq 2 \cdot 5 = 10$, $\dfrac{0}{3} \neq \dfrac{3}{b}$ since $0 \cdot b = 0 \neq 3 \cdot 3 = 9$, $\dfrac{-3}{2} = \dfrac{3}{-2}$

since $(-3)(-2) = 6 = 2 \cdot 3$, $\dfrac{-1}{-2} = \dfrac{1}{2}$ since $(-1)(2) = -2 = (-2)(1)$.   **2.** $\dfrac{a}{b} = \dfrac{ka}{kb}$

by definition of equality in F because $a \cdot kb = akb = b \cdot ka$.   **3.** $\dfrac{0}{c} = \dfrac{0}{d}$ by definition

of equality in F since $0 \cdot d = 0 = c \cdot 0$.   **4.** (a) $\dfrac{9}{8}$ (b) $\dfrac{7}{20}$ (c) $\dfrac{27}{-8}$ (d) $\dfrac{121}{-48}$

Section 8.2

1. 2520.   **2.** 7000.   **3.** 672.   **4.** 7200.   **5.** $\dfrac{5}{6}$.   **6.** $\dfrac{47}{120}$.   **7.** $\dfrac{127}{120}$.   **8.** $\dfrac{127}{120}$.

9. $\dfrac{31}{16}$.   **10.** $\dfrac{29}{30}$.   **11.** $\dfrac{31}{16}$.   **12.** $\dfrac{77}{192}$.   **13.** $\dfrac{7}{10}$.   **14.** $\dfrac{1259}{969}$.   **15.** $\dfrac{479}{140}$.   **16.** $\dfrac{87}{70}$.

17. (a) is not   (b) is.   **18.** (a) is not   (b) is not.

Section 8.3

1. (a) $\dfrac{7}{3} + \dfrac{0}{5} = \dfrac{35 + 0}{15} = \dfrac{35}{15} = \dfrac{7}{3}$; hence, by definition of subtraction, $\dfrac{7}{3} - \dfrac{0}{5} = \dfrac{7}{3}$.

(d) $\dfrac{56}{21} \times \dfrac{1}{2} = \dfrac{56}{21 \cdot 2} = \dfrac{8}{3 \cdot 2} = \dfrac{8}{6} = \dfrac{16}{12}$; hence, by definition of division, $\dfrac{16}{12} \div \dfrac{56}{21} = \dfrac{1}{2}$.

2. $\dfrac{-103}{120}$.   **3.** $\dfrac{-127}{120}$.   **4.** $\dfrac{40}{3}$.   **5.** $\dfrac{23}{10}$.   **6.** $\dfrac{56}{15}$.   **7.** $\dfrac{34}{9}$.   **8.** $\dfrac{14}{15}$.   **9.** $\dfrac{35}{24}$.

10. $\dfrac{49}{-36}$.   **11.** $0 \div \dfrac{c}{d} = \dfrac{0}{1} \div \dfrac{c}{d} = \dfrac{0}{1} \times \dfrac{d}{c} = \dfrac{0 \cdot d}{1 \cdot c} = \dfrac{0}{c} = 0$.   **12.** (a) $f = \dfrac{1}{2}$, $g = \dfrac{1}{4}$

(b) $f = \dfrac{1}{2}$, $g = \dfrac{1}{4}$.   **13.** $a = 2$, $b = 3$.

Section 8.4

Let $f = \dfrac{a}{b}, g = \dfrac{c}{d}$ and $h = \dfrac{r}{s}$ where a, c, and r are integers and b, d, and s are nonzero

integers.   **1.** $f \cdot 1 = \dfrac{a}{b} \times \dfrac{1}{1} = \dfrac{a}{b} = f$ and $1 \cdot f = \dfrac{1}{1} \cdot \dfrac{a}{b} = \dfrac{a}{b} = f$, hence $f \cdot 1 = 1 \cdot f = f$.

2. $fg = \dfrac{a}{b} \cdot \dfrac{c}{d} = \dfrac{ac}{bd}$, $gf = \dfrac{c}{d} \cdot \dfrac{a}{b} = \dfrac{ca}{db} = \dfrac{ac}{bd}$ since multiplication is commutative in I.

Hence $fg = gf$.   **3.** $fg = \dfrac{a}{b} \cdot \dfrac{c}{d} = \dfrac{ac}{db}$. Now ac and bd are in I since I is closed under

multiplication. Also $bd \neq 0$ by Theorem 6.7.2. part (2). Therefore, fg is in F.

4. $(f + g) + h = \left(\dfrac{a}{b} + \dfrac{c}{d}\right) + \dfrac{r}{s} = \dfrac{ad + bc}{bd} + \dfrac{r}{s} = \dfrac{(ad + bc)s + bdr}{bds} = \dfrac{ads + bcs + bdr}{bds}$.

$f + (g + h) = \dfrac{a}{b} + \left(\dfrac{c}{d} + \dfrac{r}{s}\right) = \dfrac{a}{b} + \dfrac{cs + dr}{ds} = \dfrac{ads + b(cs + dr)}{bds} = \dfrac{ads + bcs + bdr}{bds}$.

Therefore, $(f+g)+h=f+(g+h)$.   **5.** By Theorem 8.3.1, $f-g=\dfrac{a}{b}-\dfrac{c}{d}=$

$\dfrac{ad-bc}{bd}$. Since I is closed under multiplication and subtraction, $ad-bc$ and bd are

in I. Also $bd\neq 0$ by Theorem 6.7.2 part (2). Therefore, $f-g$ is in I.   **6.** $f(g+h)=$

$\dfrac{a}{b}\left(\dfrac{c}{d}+\dfrac{r}{s}\right)=\dfrac{a}{b}\left(\dfrac{cs+dr}{ds}\right)=\dfrac{a(cs+dr)}{bds}=\dfrac{acs+adr}{bds}.\ fg+fh=\dfrac{a}{b}\cdot\dfrac{c}{d}+\dfrac{a}{b}\cdot\dfrac{r}{s}=\dfrac{ac}{bd}+\dfrac{ar}{bs}$

$=\dfrac{abcs+bdar}{bdbs}=\dfrac{b(acs+dar)}{b(bds)}=\dfrac{acs+dar}{bds}$. Therefore, $f(g+h)=fg+fh$. To prove

that $(g+h)f=gf+hf$, uses the commutative property of multiplication (part (7)) to
get $f(g+h)=(g+h)f$ and $fg+fh=gf+hf$. But we have just proved that $f(g+h)=$
$fg+fh$, and hence by the transitive property of equality $(g+h)f=gf+hf$.
7. (a) $f=8,\ g=5,\ h=1$ (b) $f=12,\ g=6,\ h=2$.   **8.** (a) $h=24,\ f=2,\ g=6$,
(b) $h=24,\ f=2,\ g=6$.

Section 8.5

1. (a) $-\frac{7}{3}$ (b) 6 (c) $\frac{5}{2}$ (d) 0 (e) 0 (f) 0 (g) $\frac{2}{11}$ (h) $\frac{3}{7}$ (i) $\frac{2}{11}$ (j) $-\frac{3}{4}$ (k) 0
(l) $\frac{3}{5}$.   **2.** Since $\frac{2}{7}-\frac{5}{7}=-\frac{3}{7}$ a negative rational number, the positive rational
numbers are not closed under subtraction.

Section 8.6

1. $-\dfrac{7}{13}<-\dfrac{19}{38}<-\dfrac{17}{35}<0<\dfrac{1}{4}<\dfrac{8}{17}<\dfrac{17}{35}<\dfrac{21}{43}<\dfrac{1}{2}=\dfrac{19}{38}<\dfrac{7}{13}<\dfrac{5}{9}<\dfrac{4}{7}<\dfrac{2}{3}$.

2. Prove that when $h>0$, if $f<g$ then $fh<gh$. Since $f<g$ there exists a positive
rational number p such that $f+p=g$. Hence, $(f+p)h=gh$. Therefore, $fh+ph=gh$.
But both p and h are positive, so ph is positive. Thus by definition of " $<$ ", $fh<gh$.
3. If $f<g$, then by Theorem 8.6.3 part (2), $f+f<f+g$ and $f+g<g+g$. Thus
$2f<f+g<2g$ and, hence, $f<\frac{1}{2}(f+g)<g$.

Section 8.7

1. 17.   **2.** $14\frac{11}{17}$.   **3.** $-30\frac{2}{3}$.   **4.** $-257\frac{2}{5}$.   **5.** $12\frac{23}{29}$.   **6.** $-2\frac{25}{31}$.   **7.** $12\frac{23}{30}$.
8. $36\frac{7}{24}$.   **9.** $2\frac{20}{21}$.   **10.** $4\frac{2}{9}$.

Section 8.8

2. (a) Take a unit of measure that fits exactly 3 times in the interval from 0 to 1,
and move 2 of these units to the right of the origin to locate the point 2/3.   (b) As
in (a), but move 7 such units to the right of the origin.   (e) As in (a), but move 4
such units to the left of the origin to locate the point $-\frac{4}{3}$.   **3.** (a) The set of all
rational points to the left of the point $\frac{7}{8}$.   (b) The set of all rational points to the
right of the point $-\frac{1}{2}$.   (c) The set of all rational points between the point $-\frac{3}{4}$ and
the point 2.   (d) The set of all rational points between the points $\frac{7}{16}$ and $\frac{1}{8}$.   (e) The
set of all rational points that are to the left of 2 and to the right of 3. There are no
such points so the set is $\varnothing$.   (f) The set of all rational points to the left of 2 and to
the left of 1. This is the set of all rational points to the left of 1.   (g) The set of all
rational points that are either to the left of -2 or to the right of 3.   **4.** (a) 3
units   (b) $\frac{1}{14}$ unit   (c) $4\frac{6}{65}$ units   (d) $4\frac{14}{19}$ units.

5. (a)

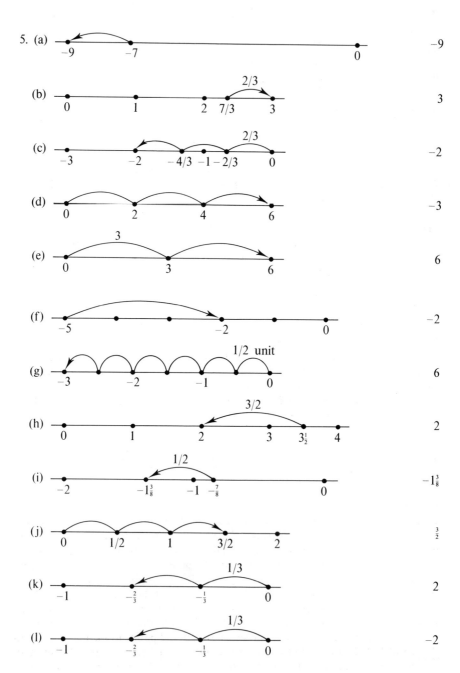

(b)

(c)

(d)

(e)

(f)

(g)

(h)

(i)

(j)

(k)

(l)

−9

3

−2

−3

6

−2

6

2

$-1\frac{3}{8}$

$\frac{3}{2}$

2

−2

Section 8.9

1. $\left(\dfrac{a}{b}\right)\left(-\dfrac{c}{d}\right) = \left(\dfrac{a}{b}\right)\left(\dfrac{-c}{d}\right) = \dfrac{a(-c)}{bd} = \dfrac{-ac}{bd} = -\dfrac{ac}{bd} = -\left(\dfrac{a}{b} \cdot \dfrac{c}{d}\right).$

2. $\dfrac{a}{b} \div \left(-\dfrac{c}{d}\right) = \dfrac{a}{b} \div \dfrac{-c}{d} = \dfrac{a}{b} \cdot \dfrac{d}{-c} = \dfrac{ad}{b(-c)} = \dfrac{ad}{-bc} = -\dfrac{ad}{bc} = -\left(\dfrac{a}{b} \div \dfrac{c}{d}\right).$ 3. $\dfrac{5}{6}.$

4. $\dfrac{1}{4}.$ 5. $-\dfrac{22}{15}.$ 6. $\dfrac{43}{14}.$ 7. $-\dfrac{51}{40}.$ 8. $\dfrac{2}{3}.$ 9. $\dfrac{14}{25}.$ 10. $-\dfrac{1}{6}.$ 11. $-\dfrac{1}{6}.$

12. $-\dfrac{6}{5}.$ 13. $\dfrac{11}{4}.$ 14. $-\dfrac{5}{8}.$ 15. $-\dfrac{25}{6}.$ 16. $\dfrac{-31}{35}.$

Section 8.10

1. (a) $\frac{3}{2}$ (b) $\frac{7}{8}$ (c) 0 (d) 0 (e) $\frac{5}{24}$.   2. (a) 0 (b) 0.   3. (a) $v - u$ (b) $v - u$.
4. (a) $u - v$ (b) $u - v$.   5. 6, -6.   6. 0.   7. 5, -5.   8. all $x \neq 0$.   9. No values.   10. 2.   11. -2.   12. No values.   13. The set of all rational points x whose distance from the origin is less than 3.

$\{x \in F \mid -3 < x < 3\}$

14. The set of all rational points x whose distance from the origin is greater than 2.

$\{x \in F \mid x < -2 \text{ or } x > 2\}$

15. The set of all rational points x whose distance from the point 2 is less than 1.

$\{x \in F \mid 1 < x < 3\}$

16. The set of all rational points x whose distance from the point $\frac{1}{2}$ is greater than 3.

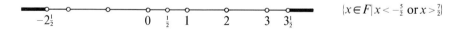

$\{x \in F \mid x < -\frac{5}{2} \text{ or } x > \frac{7}{2}\}$

17. The set of all rational points x whose distance from the point -2 is greater than 5.

$\{x \in F \mid x < -7 \text{ or } x > 3\}$

18. The set of all rational points x whose distance from the point $-\frac{2}{3}$ is less than $\frac{7}{8}$.

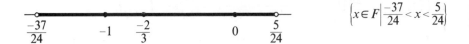

$$\left\{x \in F \,\middle|\, \frac{-37}{24} < x < \frac{5}{24}\right\}$$

19. The set of all rational points x whose distance from the point 1 is less than 0. This is $\varnothing$.   **20.** The set of all rational points x whose distance from the point 3 is less than or equal to zero.

$\{3\}$

21. The set of all rational points x whose distance from the point -3 is less than or equal to zero.

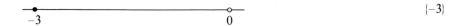

$\{-3\}$

22. The set of all rational points x whose distance from the point 3 is greater than -2.

F (all rational points)

23. The set of all rational points x whose distance from the point -2 is less than -1. This is $\varnothing$.   **24.** The set of all rational points x whose distance from the point 1 is equal to 3.

$\{-2, \quad 4\}$

25. The set of all rational points x whose distance from the point 5 is equal to 2.

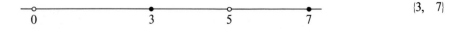

$\{3, \quad 7\}$

26. The set of all rational points x whose distance from the point -7 is equal to 4.

$\{-11, \quad -3\}$

27. The set of all rational points x whose distance from the point -1 is equal to -3. This is $\varnothing$.　　**28.** The set of all rational points x whose distance from the point -2 is less than zero. This is $\varnothing$.

Section 8.11
1. $\{3\}$.　　**2.** $\{\frac{24}{7}\}$.　　**3.** $\{x \in F \mid x > 3\}$.　　**4.** $\{x \in F \mid x > \frac{24}{7}\}$.
5. $\{x \in F \mid -15 < x < -\frac{11}{2}\}$.　　**6.** $\varnothing$.　　**7.** $\{x \in F \mid x > 5\}$.　　**8.** $\varnothing$.
9. $\{x \in F \mid x > -4\}$.　　**10.** $\{x \in F \mid -1 < x < 11\}$.　　**11.** $\{x \in F \mid -3 < x < 3\}$.
12. $\{x \in F \mid x < -2 \text{ or } x > 2\}$.　　**13.** $\{x \in F \mid -9 < x < -3\}$.
14. $\{x \in F \mid x < -7 \text{ or } x > -1\}$.　　**15.** $\{x \in F \mid -3 < x < 7\}$.　　**16.** $\{0\}$.
17. $\varnothing$.　　**18.** F.　　**19.** F.　　**20.** $\{x \in F \mid -5 < x < 5\}$.　　**21.** $\{4, 10\}$.
22. $\{-8, -4\}$.

Section 9.1

1. $\dfrac{9}{16}$.　　**2.** $\dfrac{25}{16}$.　　**3.** 1.　　**4.** 25.　　**5.** $\dfrac{16}{81}$.　　**6.** -8.　　**7.** $\dfrac{4^5}{3^5}$.　　**8.** 5^{-6}.　　**9.** $\dfrac{1}{9}$.

10. 8.　　**11.** $\dfrac{1}{64 \cdot 81} = \dfrac{1}{5184}$.　　**12.** $\dfrac{4}{9}$.　　**13.** 2^{-20}.　　**14.** $-\dfrac{81}{16}$.　　**15.** $2 \cdot 10^{-7}$.
16. $2(10)^6 + 3(10)^5 + 5(10)^4 + 7(10)^3 + 9(10)^2$.　　**17.** $3(10)^{-2} + 2(10)^{-3} + (10)^{-4}$.
18. $3(10) + 5(10)^{-2} + 6(10)^{-3}$.　　**19.** $\frac{25}{9}$.　　**20.** a^{-1}.　　**21.** $a^{-15}b^{20}$.　　**22.** $a^{-1}b^{-14}$.
23. 1.　　**24.** 1.　　**25.** 1.　　**26.** 10^4.　　**27.** $a^8 b^{12}$.

Section 9.2
1. $0.\overline{6}$.　　**2.** $0.1\overline{6}$.　　**3.** 0.125.　　**4.** 0.025.　　**5.** $1.\overline{54}$.　　**6.** $1.\overline{714285}$.　　**7.** $0.\overline{3307692}$.
8. 0.01.　　**9.** 0.005.　　**10.** 0.001.　　**11.** 0.184523809.　　**12.** $0.00\overline{3}$.　　**13.** 6.713×10^2.
14. 8.49×10^{-2}.　　**15.** 6.5×10^{-6}.　　**16.** 3.7×10^6.　　**17.** 7×10^{-2}.　　**18.** 6.801×10^4.

Section 9.3
1. 69.2674.　　**3.** -65.2946.　　**5.** 0.0358668.　　**7.** 7.30762431.　　**9.** 1600.

11. $122{,}666.\overline{6}$.　　**13.** $247.\overline{619047}$.　　**15.** $0.20 = \dfrac{1}{5}$.　　**16.** $0.\overline{3} = \dfrac{1}{3}$.　　**17.** $0.05 = \dfrac{1}{20}$.

18. $0.01 = \dfrac{1}{100}$.　　**19.** $0.0001 = \dfrac{1}{10{,}000}$.　　**20.** $0.005 = \dfrac{1}{200}$.　　**21.** $0.00002 = \dfrac{1}{50{,}000}$.

22. $1.0 = 1$.　　**23.** $1.5 = \dfrac{3}{2}$.　　**24.** $\dfrac{x}{100}$ (cannot express as a decimal without knowing x).　　**25.** 50%.　　**26.** 25%.　　**27.** $66\frac{2}{3}\%$.　　**28.** 250%.　　**29.** 1%.　　**30.** 30%.
31. 0.7%.　　**32.** 360%.　　**33.** 64%.　　**34.** 205%.

Section 9.4
1. decimal places 507.3, 507.29, 507.286, 507.2856, 507.285650; places $500.$, $510.$, $507.$, 507.3.　　**3.** decimal places 62.5, 62.50, 62.500, 62.5000, 62.499988; places $60.$, $62.$, 62.5, 62.50.　　**5.** sums 34.4, 34.43, 34.426, 34.4263, 34.4262626; differences 23.2, 23.13, 23.130, 23.1293, 23.1292930.　　**7.** sums 0, 0.03, 0.030, 0.0306, 0.0305988; differences 0, 0.01, 0.016, 0.0156, 0.0156638.　　**9.** products 5000, 4800, 4810, 4811.7; quotients 70, 73, 73.1, 73.137.　　**11.** products 200, 170, 175, 175.64; quotients $8{,}000{,}000$, $8{,}300{,}000$, $8{,}150{,}000$, $8{,}137{,}700$.

Section 9.6

2. (a) $\dfrac{25}{4}$ **(b)** $\dfrac{3}{400}$ **(c)** $\dfrac{631}{10{,}000}$ **(d)** $\dfrac{4{,}317}{999}$ **(e)** $\dfrac{61{,}106}{9900}$ **(f)** $\dfrac{35{,}883}{990{,}000}$.   **3. (a)** $5.\overline{56}$

(b) $6.2\overline{32}$ **(c)** $0.\overline{35}$ **(d)** $0.131031003100031\ldots$ where the pattern continues.

Section 9.8

1. If there were rational number h such that $h^2 = 3$, then there would be integers n and d such that $g(n, d) = 1$ and $(n/d)^2 = 3$. Then $n^2 = 3d^2$. But factorization into primes is unique, so $3 \mid n$. Therefore, $n = 3k$ for an integer k. Then $n^2 = (3k)^2 = 9k^2 = 3d^2$. Hence, $3k^2 = d^2$ and so $3 \mid d$. Thus $3 \mid n$ and $3 \mid d$ contradicting the fact that $g(n, d) = 1$. Consequently, no rational number h exists such that $h^2 = 3$ and hence $\sqrt{3}$ is not rational.   **3.** (Using $1/2''$ intervals.)

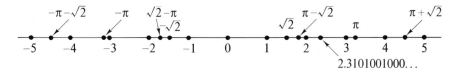

Section 9.9

1. (a) $0.13682, 0.136796, 0.136793$;   **(c)** $-4.706, -4.71, -4.711$.   **2. (a)** upper bounds 5, 6, 7, least upper bound 5, a rational number.   **(b)** upper bounds 2, 7, 35, least upper bound $\sqrt{3}$, irrational.   **(c)** upper bounds 2, 3, 4, least upper bound $\sqrt{3}$. **(d)** no upper bound exists.   **(e)** upper bounds $-2, -1, 7$, least upper bound $-\pi$, irrational.

INDEX